Adjudication under the
Sch
Cor

Adjudication under the Scheme for Construction Contracts

INCLUDING PAYMENT PROVISIONS

Second edition

Guy Cottam

Thomas Telford

Published by Thomas Telford Ltd, 1 Heron Quay, London E14 4JD.
URL: http://www.thomastelford.com

First published 1998
Second edition 2002

Distributors for Thomas Telford books are
USA: ASCE Press, 1801 Alexander Bell Drive, Reston, VA 20191-4400
Japan: Maruzen Co. Ltd, Book Department, 3–10 Nihonbashi 2-chome, Chuo-ku, Tokyo 103
Australia: DA Books and Journals, 648 Whitehorse Road, Mitcham 3132

A catalogue record for this book is available from the British Library

ISBN: 0 7277 3147 5

Typeset by Gray Publishing, Tunbridge Wells, Kent
Printed and bound in Great Britain by MPG Books Ltd, Bodmin, Cornwall

Contents

Preface

The first edition of this book proved to be very popular and was even referred to in court. However, since it was written we have now had three years of experience with this new form of statutory adjudication and the legal position of adjudication has become clearer.

Adjudication has been an outstanding success. Paying parties have realised that they now have to deal with a payee's claims quickly and efficiently and payees have learnt that they have to present their claims in a matter which reflects their contractual rights and not opportunistic whims. It has made all parties more aware of their obligations to each other.

There have now been around 200 court judgements, mostly on enforcement proceedings. There have been four decisions of the Court of Appeal. It is clear that the courts have been supportive of adjudication and are clearly of the opinion that adjudicators' decisions should be enforced summarily even if they contain mistakes unless it can be shown that the payee would be unable to repay a monetary award should the adjudicator's decision be reversed. The safeguard is that the dispute may subsequently be referred to arbitration or litigation.

The effects of these judgements have been reflected in the text.

Experience has shown that it is not necessary to have to take advice before embarking upon an adjudication. All those who are capable of negotiating their own contracts are quite capable of preparing their case for the adjudicator. It is even possible to seek enforcement of the decision through the courts without legal representation.

I have included in this edition additional guidance on how to present your information to the adjudicator, what to expect from the other side and how to enforce a decision if the other party does not implement it. I have not tried to turn the book into a legal treatise although there is comment on some of the more important cases at the end.

Experience has shown that probably the most important feature of the Housing Grants Construction & Regeneration Act 1996 has been the payment provisions. The section on the payment provisions has been expanded to take account of our better understanding of their effects. The dangers of paying parties failing to give the appropriate notices has become apparent.

Experience has also shown that parties are reluctant to pursue a dispute through the courts or arbitration if they disagree with the adjudicator's decision. This is probably due to the costs involved for what will probably be only a marginal advantage. There is little reason, generally, to be confident that a judge or arbitrator will come to a more favourable decision than the adjudicator did. It is therefore important to ensure that the best possible case is presented to the adjudicator and it is presented in a manner that will assist the adjudicator in reaching the Decision. I have tried to indicate what the adjudicator requires.

Guy Cottam

Preface to the first edition

The aim of this book is to provide a simple guide to the effects of the Housing Grants, Construction and Regeneration Act 1996, and the Scheme for Construction Contracts which is a Regulation produced under the Act.

Although the text follows in detail the Scheme for England and Wales, the Schemes for Scotland and Northern Ireland are similar with one major and a number of minor alterations. The main difference concerns the enforcement of an adjudicator's decision which in England and Wales relies on the Arbitration Act 1996, but in Scotland permits registration of the decision in the Books of Council and Session. Otherwise the Scottish version merely tidies up some of the more blatant meandering draughtsmanship of the English version, with minimal consequences. Users in Scotland should check the text of the paragraphs in the Scottish version to identify the differences, which are highlighted in Appendix C.

The Scheme implements the purposes of the Act if the parties to the construction contract fail to include its requirements in their contract. The Scheme has been drafted in parliamentary language, and is not particularly user-friendly.

The most important requirement of the Act is the right to adjudication for all disputes. Adjudication not only covers all disputes that may arise on a construction contract but is also the principal means by which the other requirements will be enforced. This is why most of the book is devoted to adjudication.

There is no reason why a party wishing to go to adjudication should employ outside help. For most disputes the parties will be able to produce all the notices and other documents required by the adjudicator themselves. The main problem will be time, particularly for the party who has not initiated the adjudication, for he will have only about three weeks to respond, whereas the initiating party may spend a considerable amount of time preparing his case before starting the adjudication process.

This book does not go through the Scheme clause by clause, but explains the actions required by the Scheme stage by stage. It is not intended to be a legal treatise but a simple guide to help those who have a dispute which is referred to adjudication, or those who have difficulty with the payment terms.

It explains to those people their new rights under the Act, and how to go about securing them. It does not cover disputes which may arise later concerning the conduct of the adjudication or interpretation of the Act itself. These types of dispute will inevitably be referred to either arbitration or litigation and further specialist advice will be needed for them.

At every stage the requirements of the Scheme are printed in italics, and the various chapters include all the words contained in the Scheme. Alternative actions are discussed at each stage, and appropriate tactics highlighted.

The Scheme applies to a wide range of construction contracts, not just to contracts between client and a contractor. The scheme also covers consultants' contracts with their clients, main contracts, subcontracts, subsubcontracts and so on down the chain. It is therefore difficult to define the names of the parties to the relevant construction contracts. Accordingly, I have referred to the party for whom the service is provided and who will be paying for the service as the employer. This term will therefore refer to the contractor in the context of a subcontract dispute.

The Act covers a wide range of contracts so it is impossible to describe every potential party to a dispute by the name that he would normally ascribe to himself. Throughout this book I have referred to the person who starts the adjudication as the Referring Party and the person who receives the notice as the Receiving Party. There is, unfortunately, not yet any universally accepted terminology, and some commentators refer to the responding party or the defending party. What we must avoid is using the terms used in litigation or arbitration. I have also used the term 'employer' for the party under the contract who is paying for the service and this will include a contractor in the case of a subcontract and a consulting engineer or architect when he engages others to do work for him.

Although this book is written using the masculine gender for simplicity of reading, it should be read as including the feminine throughout.

Guy Cottam

1

Introduction: the Act and its implications

In July 1994, Sir Michael Latham published the results of his investigation into the construction industry in a report titled **Constructing the Team.**[1] In his report, Sir Michael made a number of proposals for fairer contracts.

He recommended that the industry should be encouraged to use certain standard contracts and that these should be unamended. He said that the Bill should state that when any of the standard forms are used the following actions should be unfair or invalid:[2]

(1) any attempt to amend or delete the sections relating to times and conditions of payment, including the right of interest on late payment;
(2) to seek to deny or frustrate the right of immediate adjudication to any party to the contract or subcontract, where it has been requested by that party;
(3) to refuse to implement the decision of the adjudicator;
(4) to seek to exercise any right of set-off or contracharge without:
  (*a*) giving notice in advance;
  (*b*) specifying the exact reason for deducting the set-off; and
  (*c*) being prepared to submit immediately to adjudication and accepting the result subject to (3) above.

1 Final report published by HMSO ISBN 0 11 7552994.
2 Paragraph 8.9 & Recommendation 25: Unfair Conditions.

(5) to seek to set-off in respect of any contract other than the one in progress.

And in bespoke contracts these points should be declared unfair or invalid and in addition any attempt by contractors to introduce 'pay when paid' conditions should be explicitly declared unfair and invalid.[3]

He said that in order to aid confidence, these central provisions should be underpinned by legislation.

This prompted the Conservative Government of the day to pass **The Housing Grants, Construction and Regeneration Act 1996**[4] (The Act) which imposes some of these principles upon the construction industry.

The Government decided not to introduce these principles into legislation in a form similar to the Unfair Contract Terms Act 1977, which rendered certain terms in contract unfair, but as rights which one party is free to exercise if he so wishes.

The Act addresses three important issues in construction contracts:

(*a*) Interim settlement of disputes by adjudication;
(*b*) payment by instalments for contracts lasting longer than 45 days;
(*c*) the ability to suspend performance if not paid within a specified period; and
(*d*) the outlawing of 'pay when paid' clauses.

The Act states the basic principles which are to apply for each of these matters. It imposes them by permitting the parties to a construction contract to include them voluntarily in their contracts with such additions as they may wish, provided that those additions do not offend against the basic principles themselves.

3 Paragraph 8.10 & Recommendation 25.

4 See Appendix A.

When a construction contract does not comply with these requirements the Act provides that the **Scheme for Construction Contracts**, which is referred to in this book simply as the Scheme, applies.[5] When the Scheme applies its requirements have the effect of implied terms under the contract.[6]

To whom does the Act apply?

The Act applies to all 'construction contracts'. The Act defines what is a construction contract and the definition comes from the Income and Corporation Taxes Act 1988. It is by no means ideal and has a number of significant omissions. The Secretary of State is given power, by order, *to add to, amend or repeal any of the provisions ... as to ... which are construction contracts.*[7] This is done by laying a draft order before both Houses of Parliament.

A construction contract is an agreement with another party for any of the following:[8]

(*a*) *the carrying out of construction operations;*
(*b*) *arranging for the carrying out of construction operations by others, whether under a subcontract to him or otherwise;*
(*c*) *providing his own labour, or labour of others, for the carrying out of construction operations.*

These definitions clearly cover contracts for construction work but the Act also includes consultancy contracts in connection with construction work. Construction contracts also include an agreement:[9]

(*a*) *to do architectural, design or surveying work, or*

5 The Act ss108(5), 109(3), 110(3), 111(3), 113(6), 114(4).
6 s114(4).
7 s104(4).
8 s104(1).
9 s104(2).

(*b*) *to provide advice on building, engineering, interior or exterior decoration or on the laying-out of landscape, in relation to construction operations.*

A wide definition is provided for construction operations.[10] These include *construction, alteration, repair, maintenance, extension, demolition or dismantling of buildings, structures forming or to form, part of the land including walls, roadworks, power-lines, telecommunication apparatus, aircraft runways, docks and harbours, railways, inland waterways, pipe-lines, reservoirs, water-mains, wells, sewers, industrial plant and installations for purposes of land drainage coast protection or defence.*[11]

The definition goes on to include the incidental operations involved in constructing these works. There are, however, a number of important exclusions.

First, the construction contract must be in writing. If it is not then the Act does not apply to it. However, the definition of a contract in writing is again wide. It is said to be in writing if it is made in writing. This is the normal written agreement. However, the Act includes agreements which are evidenced in writing:[12] An agreement is evidenced in writing if the agreement is *recorded by one of the parties or a third party with the authority of the parties to the agreement.*[13] This means that if someone at a meeting for instance is asked to produce minutes of the meeting and the minutes record the agreement then it is an agreement in writing and the contract, if a construction contract, is covered by the Act. However, the court of appeal has said that the relevant terms must be in writing even if the contract itself is not in writing.[15]

The Act does not apply *to a construction contract with a residential occupier.*[14] This exemption will include houses and flats which are

10 s105.
11 s105(b).
12 s107(2)(c).
13 s107(4).
14 s106(1)(a).
15 *RJT Consulting Engineering Ltd* v. *DM Engineering (Northern Ireland) Ltd* CA (8 March 2002).

either occupied by, or are intended to be occupied by one of the parties to the contract. It does not exclude housing developments by either property companies or housing associations.

Also exempt are material and component supply contracts, mineral extraction contracts and process engineering contracts.[16]

Even when the primary operation is exempt, such as a process plant, the subsidiary operations such as site preparation, roadworks and office buildings will not be exempt. This means that those with exempt contracts will have to consider whether or not it is in their own interest to include the requirements of the Act into their contracts so as to be 'back to back' with subcontracts that they may subsequently enter into and which will be covered by the Act.

It follows that the Act will apply to most contracts in the industry: client/consultant; client/main contractor; main contractor/subcontractor; and subcontractor/subsubcontractor. It was even queried whether legal advice from a solicitor in relation to construction operations would be caught!

Before either entering into a contract or before trying to impose the requirements of the Act it would be wise to check whether or not the contract will be affected by the Act. If in doubt, it would be best to assume that it is.

Adjudication

The first of the requirements of the Act to be considered is the requirement of adjudication. This was the major recommendation of Sir Michael Latham.

The Act says that *a party to a construction contract has the right to refer a dispute arising under the contract for adjudication under a procedure complying with*[17] *this section* [of the Act] ... *at any time.*[18]

Adjudication is a procedure whereby an impartial person is asked to give a quick temporary solution to a perceived wrong.

16 s105(2).
17 s108(1).
18 s108(2)(a).

The adjudicator's decision *is binding until the dispute is finally determined by legal proceedings, by arbitration (if the contract provides for arbitration or the parties otherwise agree to arbitration) or by agreement.*[19]

The adjudicator therefore gives a decision which the parties to the contract are bound to implement but if either party is dissatisfied with the decision he can refer the dispute for a final decision by the traditional legal processes, but implementation may not be suspended pending the outcome of those proceedings.

The aim of adjudication is to resolve disputes on a temporary basis so that the work may proceed unimpeded. Certain events which commonly arise can cause a serious problem for one of the parties, and if not resolved could be disastrous.

Such events may be:

(*a*) failure to pay a sum of money;
(*b*) disagreement about a certificate, an instruction; or the quality of the design or workmanship;
(*c*) disagreement as to whether an instruction should be given;
(*d*) disagreement as to whether the work is physically or legally impossible to perform.

The adjudicator makes a decision which settles the dispute temporarily so that the work may proceed with less likelihood of serious injustice being caused. If either party disagrees with the adjudicator's decision he may refer the dispute for final determination by either an arbitrator, if arbitration has either been agreed in the contract or is subsequently agreed to by the parties, or otherwise by a court.

The parties are obliged to implement the adjudicator's decision even if the decision is not accepted by one of the parties. It is a quick process, generally completed within 28 days; although the time may be increased by agreement.

---

19 s108(3).

This means that the information given to the adjudicator may be incomplete, but the safeguard for the parties is the right to refer the dispute for reconsideration by an arbitrator or court.

How does this affect you?

The Act gives you a 'Right' to adjudication in accordance with the eight basic principles. You may be able to achieve this right in your contract by writing the adjudication clause so that it includes these eight principles. If you do not have such a clause in your contract then you exercise your right by requiring adjudication in accordance with the Scheme.

In order to satisfy this statutory 'right' the adjudication clause in the contract must:[20]

(*i*) *Enable the parties to give notice of adjudication at any time.*

(*ii*) *Provide a timetable with the object of securing the appointment of the adjudicator and referral of the dispute to him within 7 days of such notice.*

(*iii*) *Require the adjudicator to reach a decision within 28 days of referral or such longer period as is agreed by the parties after the dispute has been referred.*

(*iv*) *Allow the adjudicator to extend the period of 28 days by up to 14 days with the consent of the party by whom the dispute was referred.*

(*v*) *Impose a duty on the adjudicator to act impartially.*

(*vi*) *Enable the adjudicator to take the initiative in ascertaining the facts and the law.*

(*vii*) *Provide that the decision of the adjudicator is binding until the dispute is finally determined by legal proceedings, by arbitration (if the contract provides for or the parties otherwise agree arbitration) or by agreement.*

(*viii*) *Provide that the adjudicator is not liable for anything done or omitted in the discharge or purported discharge of his functions as*

20 s108 (2)–(4).

> *adjudicator unless the act or omission is in bad faith, and that any employee or agent of the adjudicator is similarly protected from liability.*

It should be noted that a day means a calendar day unless otherwise stated so the periods include the weekends but not Bank Holidays which are expressly excluded by the Act.[21]

The eight points are referred to as the 'eight compliance points' and if any one of these requirements is not in the clause, it does not mean that the clause is invalid or that an adjudication cannot be given in accordance with it. It merely means that either party has the 'right' to give notice to the other that he requires the dispute to be adjudicated in accordance with the Scheme.

It is possible for two adjudications to be proceeding simultaneously: one in accordance with the contract procedure and one in accordance with the Scheme for Construction Contracts. This would clearly be absurd, so if it does occur, the parties should agree to stop the adjudication under the contract clause in favour of the one under the Scheme.

There is, however, another problem of which you need to be aware. The Scheme gives a right to adjudication only on disputes arising *under the contract.* Some disputes may arise from a cause which does not arise under the contract but arises from the common law or statute law (e.g. breaches of health and safety regulations, misrepresentation or tortious claims, such as negligence, trespass or nuisance). These disputes arise *in connection with* the contract and not under it.

Traditionally most dispute clauses use the words ... **arising out of or in connection with** ... in order to avoid this difficulty. The Act does not. So a careful check needs to be made that the dispute arises under the contract because if it does not there are two possible consequences.

---

21 s116.

(*a*) A non-compliant adjudication clause in the contract would be effective to settle such a dispute if the contract included the words 'in connection with'.
(*b*) There is no right to adjudication under the Scheme.

Some currently drafted Adjudication Procedures apply to both types of dispute[22] and others do not.[23]

Decision-making

Before getting into the procedure itself, it is necessary to understand how the adjudicator will set about the task of making the decision. This is necessary so that the notices and statements can be prepared in the best manner for the adjudicator. If you give him the information in the form in which he wants it, you stand a better chance of obtaining a decision in your favour.

The adjudicator is required to decide the issues in accordance with the contract. This means that he has to approach the task in a judicial manner and adjudicators will be expected to set about their task judicially.

To do this he will need to determine the relevant facts concerning the issues. He will have to determine the relevant law and apply the relevant law to the relevant facts.

The facts surrounding a dispute are often themselves in dispute. The claiming party may say that something was agreed at a meeting and the other party may say it was not. Whether agreement was reached or not is a fact. The adjudicator would then have to decide whether an agreement was reached or not.

This may not be easy when there is a direct clash of evidence. This is where the rules come in to assist him.

22 The TeCSA Rules.
23 The ICE Adjudication Procedure (1997) normally only includes disputes arising under the contract. The Construction Industry Council's Model Adjudication Rules.

Proving your Case

The first principle is that *he who alleges must prove*. It is the person alleging a fact who has to prove it. He is said to have *the burden of proof*. Sometimes it is quite difficult to decide who has the burden.

In most cases it will be the Referring Party who is making the claim and he will have to prove his allegations. The roles are reversed when there is a counter-claim.

But what is meant by proof? When is an allegation proved?

There are two standards of proof recognised by the English law. The first is the criminal standard where the prosecution has to prove *beyond reasonable doubt* that the defendant has done whatever it is of which he stands accused. This is a higher standard than that required in a civil case.

In a civil case the standard is that the allegation has to be proved on *the balance of probabilities*. This is a lower standard. To judge it the adjudicator will ask himself the question: 'is it more likely than not that what is alleged happened?' If the answer to that question is yes then the issue is proved on the balance of probability. On expert or technical matters it may come down to which of the experts' evidence the adjudicator prefers.

The difference was exemplified in the trial in the USA of O.J. Simpson who was found not guilty of murdering his wife in the criminal courts using the standard of proof of beyond reasonable doubt. However, using the balance of probabilities it was found, in the action for damages by his wife's family in the civil courts, that it was more likely than not that he did do it.

On account of the short period of time that is allowed for making the decision, adjudicators will not be able to go into the issues to the same degree of detail as a judge or arbitrator would, but there are certain principles that are required of a judicial process, and an adjudicator may be expected to use them.

The adjudicator will not be able to enquire into and check every detail of an allegation. Consequently, the adjudicator will probably resort to another rule that judges and arbitrators often use, i.e. that if a fact alleged in one party's statement is not denied by the other it is said to be accepted.

What is not denied is taken as being accepted means that great care needs to be taken when replying to a Referring Party's statement and by the Referring Party when responding to the defence.

The cardinal rule, therefore, for all correspondence during an adjudication is to take care that any statement from the other side with which you disagree is expressly denied.

This is particularly important when the whole basis of the claim is refuted. What often happens is that the Receiving Party explains at great length why the claimant has no claim, and then ignores the figures contained in the claim, thinking that they are irrelevant. If the adjudicator then comes to the conclusion that there is justification for the claim and he then assesses its value he has only the claimant's figures to go on. He may have little option but to grant the claim in full.

Practitioners, but not lawyers, often think that to comment on the figures in a claim which they deny will somehow detract from that denial. It need not and an adjudicator will not be influenced adversely if a comment such as 'we consider this claim to be entirely unjustified for the reasons given; however, should you find that we are wrong in this we wish to make the following comments on the evaluation of the claim', introduces the comment.

Evidence

All statements should be backed up by written evidence if possible.

Where reliance is put on oral statements, a note from the person making it should be included and attached to the statement.

General statements about what happened from people who were not there are almost valueless, as are statements made second hand such as, 'Mr Jones told me that he had been told by Mr Davies that this is what happened'. It would be necessary to obtain a written statement from Mr Davies.

Although all statements require back-up, unnecessary documents should not be sent. For instance, the adjudicator will not require the whole contract but only extracts of the terms that each party thinks are relevant. If the full documents are sent the adjudicator will not read all the contract to see if there is anything relevant that he has not been told. He will assume that all that is relevant will have been drawn to his attention by the parties.

The adjudicator has no duty to assist the parties to prepare their cases or to protect them from their own folly. Indeed, he would be acting unfairly if he did.

Although the adjudicator has the authority to take the initiative to ascertain the facts and the law, most adjudicators will use this power sparingly and will only investigate matters which appear unclear from the parties' statements, and which they consider essential for reaching a decision.

If you send reports, test results or drawings to the adjudicator, mark the relevant parts to which you wish to draw his attention.

Always remember – make it easy for the adjudicator to understand all that you are trying to tell him.

Enforcement of the decision of the adjudicator

As you go through this book you will see repeated references to the enforcement of the adjudicator's decision. This is because if one party does not implement the decision of the adjudicator the other party will want to do something about it, and ultimately the only means of enforcement is the courts.

This does not mean that it is expected that the adjudicator's decisions will not be implemented. Quite the contrary. They normally will be. But there will be the odd occasion when a party is quite unwilling to do so.

The power of the court is limited to two options. It can either enforce the decision or set it aside, i.e. refuse to enforce it. The court has no power to either change the decision or remit it to the adjudicator for further consideration. If the court refuses to enforce the decision then the dispute has to be considered afresh by either the legal process or arbitration. This will take many months and may have serious consequences for one of the parties or the carrying-out of the construction works, or both.

The problem arises because Parliament when it passed the Act did not consider the subject of enforcement. They left it to be implemented in the same way as other breaches of contract.

The actual means of enforcement is outside the scope of this book and in most cases legal advice will be necessary for anyone who is unfortunate enough to experience a refusal to comply with an adjudicator's decision.

Private individuals may pursue a claim through the courts themselves; however, this is not desirable in any court other than the small claims court. Limited companies can only be represented by employees in the small claims court; in other courts they must be represented by a lawyer.

It is therefore necessary to bear in mind the arguments that a reluctant party would put to a judge in order to persuade the judge not to issue an order that implements the decision so that such arguments may be presented to the adjudicator. He can then deal with them, if he considers it necessary, in his decision.

It is also important for the adjudicator to ensure, as far as he can, that the courts will uphold his decision. To do this he should:

(*a*) follow any procedure under which he is appointed;
(*b*) only consider, without further agreement between the parties, the issues referred to in the Notice of Adjudication;
(*c*) act fairly between the parties;
(*d*) give each party a reasonable opportunity of putting its case.

Payment

The Act implements some of Sir Michael Latham's proposals with respect to a Modern Contract[24] and unfair conditions. These include some of the proposals on payment.

24 'Constructing the Team' recommendations 5.18, 8.9 & 8.10.

The Act requires the following:

(*a*) for all contracts in excess of 45 days, stage payments should be made;
(*b*) All contracts should have an adequate mechanism for deciding what payments are due and when;
(*c*) If payment is not made within a specified time performance of the contract may be suspended;
(*d*) A notice must first be given if it is intended to withhold payment (set off); and
(*e*) 'Pay when paid' clauses will be unenforceable.

Similarly to the adjudication provisions, the payment provisions in the Act in some ways go further than Sir Michael. He did not propose that there should be a right to suspension for non-payments but a right to interest at a sufficiently high rate to act as a deterrent.

Other than outlawing 'pay when paid' clauses in construction contracts, the only effect of these elaborate and complex provisions is to provide a trigger mechanism for the right to suspend performance in default of payment.

However, an adjudicator can decide what payment should have been made in any stage payment and direct that it be paid.

Stage payments

First, the Act provides that *a party to a construction contract is entitled to payment by instalments, stage payments, or other periodic payments for any work under a contract unless*

(*a*) *it is specified in the contract that the duration of the work is to be less than 45 days, or*
(*b*) *it is agreed between the parties that the duration of the work is estimated to be less than 45 days.*[25]

It is not thought that artificial agreement to a contract period of less than 45 days or a series of individual contracts each less than 45 days but to run consecutively would comply with the Act.

25 s109(1).

The parties are free to agree the amount of the payment and the intervals at which, or the circumstances in which, they become due.[26]

The section is drafted to permit payments by valuation on a measured basis or by events such as milestones. This gives wide scope to those who are drafting contracts to avoid the spirit of the Act. It would be possible to define a down payment and then intervals of say 12 months, and if the contract were only 13 months long it would effectively evade the Act.

Adequate mechanism for valuing payments

Secondly, the Act requires that *every construction contract shall provide an adequate mechanism for determining what payments become due under the contract and when.*[27]

The Act does not give any indication as to what an adequate mechanism is. Clearly the mechanism provided by the standard construction contracts would be satisfactory. The minimum requirement must be some procedure which produces a sum of money to be paid. This could be by:

(*a*) valuation by measurement of work done at specified times;
(*b*) lump sums payable when certain activities, operations, milestones or targets have been met;
(*c*) percentage amounts of the contract sum at various times;
(*d*) monthly lump sum increments;
(*e*) payments on the basis of cost expended.

The only criterion must be that the selected method is capable of giving a certain result.

The contract must also provide for when each instalment or payment becomes due. This too will require a procedure which allows a specific date to be identified as the date when payment is due. This could be

26 s109(2).
27 s110(1)(a).

(*a*) The last date of each interval.
(*b*) A specified number of days after the end of each interval.
(*c*) A specified number of days after any milestone or target is reached.
(*d*) A specified number of days after the receiving party presents his invoice.

This date must be clearly identifiable as it triggers the procedure for suspension of performance.

Dates when payments are due, and the consequence of late payment

The Act gives a party to a construction contract a right to suspend performance of his obligations if he is not paid by a specified time. This is a new right. Generally under English law, although it is a breach of contract not to pay by the due date, the remedy is interest on the overdue amount if the contract so provides. Governments and the courts have consistently rejected the automatic right to interest on overdue payments. The courts have held that the failure to pay merely caused the creditor the loss of the sum of money not paid.[28] A further rule states that debts do not usually carry interest,[29] this has now been changed by the Late Payment of Commercial Debts (Interest) Act 1998, which entitles a debtor to a rate of interest of 8% over basic rate.[30]

The law does not permit a person to suspend performance unless again the contract specifically permits it, because such an action would be a breach of contract for which the normal remedy is damages.

The right to suspend performance for late payment is therefore a radical change to English law.

The mechanism for suspension is that *the contract must provide for a final date for a payment in relation to any sum which becomes due.*[31]

28 *Williams* v. *Renolds* (1865), *Hopkins* v. *Norcross plc* (1993) 1 All ER 565.
29 *London, Chatham and Dover Rly Co* v. *South Eastern Rly Co* (1893) AC 429.
30 Now applies to all contracts entered into after 1 August 2002. Late Payment of Commercial Debts (Commencement No. 3) Order 2002.
31 s110(b).

This final date for payment is a bit of a misnomer. It is actually the date upon which the receiving party is able to give seven days' notice that, unless he is paid in full, he will suspend performance.

Where a sum due under a construction contract is not paid in full by the final date for payment and no effective notice to withhold payment has been given, the person to whom the sum is due has the right (without prejudice to any other right or remedy) to suspend performance of his obligation under the contract to the party by whom payment ought to have been made.[32]

This right may not be exercised without first giving the party in default at least seven days' notice of intention to suspend performance, stating the ground or grounds on which it is intended to suspend performance.[33]

It is important that this notice should correspond exactly with these requirements because the suspension of performance is a very serious act. A party exercising this right must be very sure of his grounds for so doing because if he is wrong he will have to pay compensation for any loss suffered by the other party.

The right to suspend performance ceases when the party in default makes payment in full of the amount due.[34]

A period during which performance is suspended in accordance with the Act extends any time for completion that may be specified in the contract,[35] but the Act does not provide for any of the additional costs that the suspension may cause the suspending party to be recovered from the party in default; nor does it provide for the recovery of any overhead costs incurred during the period of suspension.

These are matters which should be covered in the construction contract itself.

32 s112(1).
33 s112(2).
34 s112(3).
35 s112(4).

Set-off

One of the major wrongs identified by Sir Michael Latham was the indiscriminate use of set-off to avoid making payment.

Set-off is the right to deduct from monies owed by one party sums of money which that party claims it is owed by the other. Such sums may arise from many causes:

(*a*) Bad or incomplete work, where the paying party deducts sums to reflect the reduced value of the work due to the defects; this is known as abatement.

(*b*) Cross claims for damages for breach of contract where the paying party has incurred additional costs due to the breach of contract of the other; this is set-off.

The most frequent is the claim by a contractor to have been delayed by a subcontractor and consequently he deducts from monies due to the subcontractor costs that he claims to have incurred as a result of the subcontractor delays.

The Act does not stop this procedure. It does, however, require the paying party to notify the other party of the amount he is going to pay and how that sum is made up. Provided that this notice is given in the appropriate form, a party is then permitted to withhold such sums. The payee may challenge the decision to withhold money by referring it to adjudication.

Every construction contract shall provide for the giving of a notice by a party not less than 5 days after the date on which the payment becomes due from him under the contract, or would have become due if:

(*a*) *the other party had carried out his obligations under the contract, and*

(*b*) *no set-off or abatement was permitted by reference to any sum claimed to be due under one or more other contracts, specifying the amount (if any) of the payment made or proposed to be made and the basis on which that amount was calculated.*[36]

The paying party is therefore required to detail exactly how his payment is made up. It may require him to value the work done in

36 s110(2).

accordance with the contract, or to define what the value of that work is. If he is not paying the full amount he will have to say why he is not paying that amount.

This again only really matters if payment is not made by the final date for payment.

A *party to a construction contract may not withhold payment after the final date for payment of a sum due under the contract unless he has given an effective notice of intention to withhold payment.*[37]

The notice given above may suffice provided it contains sufficient information. *To be effective such a notice must specify:*[38]

(*a*) *the amount proposed to be withheld and the ground for withholding payment, or*
(*b*) *if there is more than one ground, each ground and the amount attributable to it.*

The notice must also be given a prescribed time before the final date for payment and the parties are free to agree what that period should be.[39]

If the matter is referred to adjudication and the adjudicator decides that the whole or part of the sum set off should be paid, then payment becomes due seven days after the decision or the final date for payment of that sum whichever is the later.[40]

Pay when paid clauses

The Act makes ineffective any clause in a construction contract that makes payment conditional on the payer receiving payment from a third party unless that third party is insolvent.[41]

Subcontracts should always include a provision which grants the contractor relief from paying a subcontractor if he is not paid by the employer because the employer has become insolvent. The Scheme

37 s111(1).
38 s111(2).
39 ss111(2) & (3).
40 s111(4).
41 s113(1).

does not make this exception, and when the Scheme applies, payment will have to be made by the contractor to the sub-contractor even when the employer has become insolvent.

Consequences

If the construction contract does not comply with these provisions the Scheme for Construction Contracts applies. The Scheme has been laid before both Houses of Parliament in a document entitled CONSTRUCTION CONTRACTS *The Scheme for Construction Contracts Regulations*, 1998, No. 846. A copy of it is to be found in Appendix B.

The later chapters in this book advise on what you need to do when your particular construction contract does not comply with the Act, and you have to rely upon the Scheme to exercise your new rights.

Summary

- Check whether the contract is a construction contract under s104.
- If in doubt, assume it is.
- If it does not, decide whether you wish the contract to comply with the Act.
- Check whether the contract
  - has an adjudication clause which complies with the eight compliance points
  - is over 45 days' duration, and
  - if so, check
    - whether it has provision for stage payments
    - whether there is an adequate mechanism for evaluating payments.
- Check whether the contract states when payments are due, and defines a final date for payment.
- Check whether there is a provision for giving notice of the amount to be paid.
- Check whether there is provision for an effective notice for withholding payment.
- Check whether there is a pay when paid clause.
- ARE YOU RELIANT ON THE SCHEME?

2

Pre-contract considerations

The requirements of the Act must be considered from the moment that the employer starts to think about what conditions to select for his construction contract. During this process there are a number of points that should be borne in mind as a result of the Act. These are:

(*a*) Adjudication.
(*b*) Payment by instalments.
(*c*) The mechanism for determining the amounts due.
(*d*) The dates when payments become due and the final date for payment.

In all these matters the employer must decide whether he wishes to ensure that his contract complies with the requirements of the Act, or whether he will rely on the Scheme.

There are pros and cons inherent in both methods.

All these matters may be considered separately. For instance, it is perfectly possible to have a compliant adjudication procedure in the contract and to rely on the Scheme for the payment provisions.

Adjudication

The adjudication provisions in the Scheme are not well drafted. There are a number of points which could lead to difficulty later. These are described in other chapters.

There are any number of published adjudication procedures which have been expressly drafted to comply with the Act, and a clause incorporating one of these procedures into the contract should satisfy the requirements of the Act. These procedures are listed in Appendix C.

It is suggested that a simple clause such as:

> 'Should a dispute or difference arise under the contract [or in connection with the contract] either party may serve on the other a Notice of the Adjudication requiring the dispute or difference to be referred to adjudication.
>
> The adjudication shall be conducted in accordance with the version of the adjudication procedure published by the ... (insert name of publishing body)[1] current at the date of issue of the Notice of Adjudication.'

A check should be made with the body publishing the procedure to ensure that the procedure does comply with the Act.

A copy of the procedure should be obtained and a check made to see whether there are any specified requirements for the Notice of Adjudication. If there are there should be an additional clause stating:

> 'The Notice of Adjudication should contain the following information: ... '

The requirements should then be listed.

This is so that neither party may later challenge the validity of the adjudication on the grounds that the Notice of Adjudication did not comply with the requirements of the procedure.

It will then be necessary to decide how the adjudicator is to be appointed. The adjudicator may be:

(*a*) named in the Contract;
(*b*) agreed between the parties after the contract has been formed but prior to a dispute arising;
(*c*) agreed between the parties after a dispute has arisen; or
(*d*) appointed by a nominating body.

1 For example, The Institution of Civil Engineers.

If so,
should the body be named in the contract, or
selected by the party issuing the Notice of Adjudication?

Naming the adjudicator in the contract

The advantage of naming the adjudicator in the contract is that both parties know who the adjudicator will be before a dispute arises. When a dispute arises, the named person can be contacted immediately and asked to confirm that he is able and willing to act. This will undoubtedly allow the adjudication to proceed as quickly as possible once a dispute has arisen.

It is also possible for the parties to agree with the chosen person, before a dispute arises, what his terms and conditions will be if he is called upon. This too removes a source of contention which can delay the start of an adjudication.

The disadvantage is that the chosen person may not have the required expertise for the actual dispute when it arises. However, this may not be the disadvantage that it might at first seem. Traditionally, architects and engineers have been expected to perform the adjudication function for all disputes arising under a construction contract and there are a number of people available who would be capable of deciding most disputes.

A greater disadvantage is one of perception. When an adjudicator is named in the contract the selection has generally been made by the employer, with the other party often having little say in the choice. The perception of the other party from the start, therefore, is that the adjudicator is biased in favour of the employer.

For adjudication to work successfully, it is essential that the adjudicator is acceptable to both parties.

A better procedure for the selection of the adjudicator is for the employer to select, say, three names and to offer them to the other party to choose from. If the other party for any reason cannot accept any of them, he should be able to propose a further list of names for the employer's choice. Normally, a person can be agreed during the tender period.

A further advantage of naming a person is that that person can also be named as the adjudicator in all the subcontracts that follow. This allows disputes which arise under the subcontract that cause a dispute under the main contract to be decided at the same time, a process known as joinder. This ensures that a similar decision is made in each dispute. For major disputes this is essential otherwise the adjudication process could exacerbate the problem. For instance, if a dispute arose concerning the quality of some work and one adjudicator decided that it complied with the specification and the other decided that it did not, then nothing would have been solved by the adjudication and a new problem of payment would have been created.

The JCT Contracts with adjudication provisions all have provision for naming the adjudicator in the Contract, and so does the Engineering and Construction Contract (NEC).

It is unlikely that the joining of disputes will be possible if the adjudicator is not named in the Contract on account of the short time periods imposed.

After the contract has been formed, the parties can decide whether or not to enter into a contract with the adjudicator before a dispute arises so that they will have agreed all the terms and conditions under which the adjudicator will act should he be required to do so. This is the procedure adopted by the Engineering and Construction Contract.

The adjudicator may require a retainer if he is to sign such a contract.

Agreed between the parties before a dispute arises

This has many advantages similar to naming the adjudicator in the contract, but suffers from the disadvantage that there is no real incentive for the parties actually to agree.

It is probably better to agree the adjudicator in the contract.

Agreed when a dispute arises

This might at first sight be the best arrangement. The problem is that the timetable set out by the Act makes agreement all but impossible.

The Act requires that the appointment of the adjudicator and referral of the dispute to him should be within seven days of the notification of a dispute. This period includes Saturday and Sunday. Appointing bodies are expected to make the appointment in four days. This leaves only three days for agreement, of which two may be a weekend.

Normally, when a dispute arises the receiving party is not particularly interested in a speedy resolution of the dispute and may not be willing to agree a name quickly, in which case the appointment will almost certainly be by a nominating body.

This does not mean that agreement should not be sought. Clearly it should, but the pre-contract consideration by the employer should be that, in all probability, it will be the other party who will be issuing the Notice of Adjudication, and it will be that party who will choose a nominating body if there is not agreement on either the name of an adjudicator or the name of an appointing body.

If the employer is to have a say in the selection of the appointing body for most disputes, it follows that the body should be named in the contract.

This leads on to the last method of appointment.

Appointed by a Nominating Body

This is probably the least satisfactory method of appointment and likely to be the most common.

An Adjudicator Nominating Body (ANB) is defined in the Scheme as *a body (not being a natural person and not being a party to the dispute) which holds itself out publicly as a body which will select an adjudicator when requested to do so by a referring party.*[2] Originally it had been intended that there should be some form of quality control over ANBs but that is no longer so.

It is for the Issuing Party to choose an ANB if no appointing body is named in the contract and he may select a body wholly

2 Paragraph 2(3).

unacceptable to the receiving party. There is almost no possibility of the receiving party objecting either to the ANB or the person whom it appoints as adjudicator.

The parties should, wherever possible, name an appointing body in the contract. The parties will have little or no influence in respect of who is selected.

The choice of nominating body is important. Some bodies will undoubtedly have higher standards of training and selection than others.

Most of the ANBs are professional institutions which are identifiable with a particular branch of the industry: the Institution of Civil Engineers (ICE), the Royal Institution of Chartered Surveyors (RICS), the Royal Institute of British Architects (RIBA) or the Chartered Institute of Building (CIOB). Others such as the Chartered Institute of Arbitrators are more general bodies. Normally, the choice may suggest itself according to the aim of the contract, but before finally selecting an ANB the parties should enquire into what qualifications are required before a person is permitted entry to the body's list of adjudicators.

The requirements to look for are that the prospective adjudicator:

(*a*) has had at least 10 years' experience in his trade or calling;
(*b*) has obtained a knowledge of the law by some recognised route;
(*c*) has attended at least three days' adjudicator training;
(*d*) has wide experience in dispute resolution;
(*e*) has attended an interview.

Once he is on the list, he should be required to undertake a specified routine of Continuing Professional Development (CPD).

Payment provisions

The Act permits a fair degree of freedom in the application of the payment provisions, with the exception of the *pay when paid*[3] provision.

3 s113 of the Act and The Scheme – Part II Payment paragraph 13.

To comply with the Act the contract has to contain the following provisions and the Scheme provides a default in each case.

(*a*) There must be Stage payments for contracts in excess of 45 days; and if no intervals for instalments are specified in the contract the period under the Scheme is 28 days.

(*b*) There must be an adequate mechanism for deciding what payments are due and when.

If there is no adequate mechanism for determining the amount due, the Scheme includes a mechanism which is broadly in line with normal practice. The valuation will consist of:

(*i*) the value of the work done to the end of the relevant period;

(*ii*) the value of materials delivered to site;

(*iii*) any other amounts specified in the contract; and

(*iv*) previous payments.

If no date for payment of the instalment is specified in the contract the sum assessed in (*b*) above becomes due 7 days after the end of the relevant period or on presentation of the claim by the payee, whichever is the later.

(*c*) There must be a right to suspend performance if payment is not made within a specified time. The Scheme provides what is known as the final date for payment after which a party is entitled to give 7 days' notice of suspension of performance. The final date for payment is 17 days[4] after the sum becomes due.

(*d*) A notice must be given by the paying party stating the amount that will be paid in each instalment. This notice of how much will be paid must be given under the Scheme no later than 5 days after the sum became due.

(*e*) A notice must first be given if the paying party is intending to withhold payment from the other party. The sums withheld are generally in respect of claims that the paying party alleges he has against the other party. This is the process known as set-off. The

---

4 See Figure 2.

Act permits such set-off provided that a notice is given detailing how the amounts withheld have been made up. If no time for giving this notice is specified in the contract the Scheme requires that such a notice be given within 10 days[5] after the sum became due.

(*f*) The Act makes any clause in a contract which makes payment dependent upon the payer receiving payment from a third party unenforceable. If a contract has a 'pay when paid' clause in it then that clause is ineffective, and unless there are other terms of payment in the contract (which there usually will not be) the terms for payment detailed in the Scheme, as detailed above, will apply.

For most of these time periods the Act permits the parties to agree what they shall be. The Scheme provides what in practice will be minimum periods. Employers are going to find that the specified periods may be too short for their internal procedures and in the case of suspension of work too short to be acceptable.

In that case, they should draft their own terms of payments, making sure that they comply with the requirements of the Act.

5 See Figure 2.

3

Adjudication — getting started: the Notice of Adjudication

The construction industry actually has comparatively few disputes. There are many arguments but only a small number are referred to arbitration[1] or the courts, and of those, only about 15% do not settle before the trial.

Most parties are able to settle their differences, but when that proves impossible and one party is not satisfied with his current position he will have to decide whether to accept the position or start the dispute settlement procedure.

The Act does not require a party to go to adjudication before a dispute is referred to arbitration or the courts. Adjudication is a right that the parties may invoke if their contract does not include an adjudication clause.

When a settlement cannot be reached the aggrieved party must decide whether or not he wishes to invoke arbitration immediately. He will, of course, have to if the contract requires it, but if it does not he will have to decide whether or not he wants an adjudication under the Scheme.

Whether the adjudication is under a contractual clause or under the Scheme, the initial action is likely to be the same — the service of a Notice of Adjudication upon the other party.

---

1 It was estimated that about 1000 arbitrators are appointed each year for construction disputes prior to the Act. The current figure is nearer 300.

This Notice is important because it defines the matters in dispute and therefore the matters that are being referred to the adjudicator. Unless the adjudication procedure under which the adjudication is to be carried out provides expressly for additional disputes to be added the adjudicator cannot consider issues which are not included in the Notice of Adjudication.

Starting the Adjudication under the Scheme

The Scheme requires a written notice, the Notice of Adjudication, to start the adjudication procedure. This must be sent *to every other party to the contract.*[2]

Any party to a construction contract ('the referring party') may give written notice (the 'Notice of Adjudication') of his intention to refer any dispute arising under the contract, to adjudication.[3]

The Notice of Adjudication serves four purposes.

(*a*) It informs the other parties to the contract that there is a dispute and the nature of that dispute so that they can consider either reaching a settlement or preparing their defence for the adjudicator.
(*b*) It informs the adjudicator of the nature of the disputes so he can decide whether he is suitably qualified to decide the issues.
(*c*) It provides the basic information required by either an appointing body or an ANB for selecting an appropriate person to act as adjudicator.
(*d*) It defines the jurisdiction (the matters that he is empowered to consider) of the adjudicator.

Anyone drafting the notice should bear these purposes in mind.

The Scheme requires specific information. It says the notice should state briefly:[4]

(*a*) *the nature and a brief description of the dispute and of the parties involved, and*

---

2 Paragraph 1(2) of the Scheme for Construction Contracts.
3 Paragraph 1(1).
4 Paragraph 1(3) (a)–(d).

(*b*) *details of where and when it has arisen;*
(*c*) *the nature of the redress sought, and*
(*d*) *the names and addresses of the parties to the contract (including, where appropriate, the addresses which the parties have specified for the giving of notices).*

The nature of the dispute and brief description should be fairly general so as not to restrict the issues unduly. However, because the appointing authorities do not know the type of work being executed under the contract it will be necessary to state what it is. For instance, 'the dispute concerns the valuation of various variation orders issued by the project manager between 1/1/98 and 3/4/98' would not be very helpful, but the mere addition of the words 'concerning ground works' would add substantially to its usefulness.

Of particular importance will be:

(*a*) the category of work: building, civil engineering, process engineering or mechanical and electrical work;
(*b*) the type of work: ground work, tunnelling, brickwork, concrete (reinforced or prestressed), finishes, plumbing, heating and ventilating, lifts, etc.;
(*c*) legal: interpretation of the contract, meaning of the specification, misrepresentation, tortious claims;
(*d*) whether technical, workmanship, compliance, or valuation.

A brief description of how the dispute arose would be helpful, as it will probably include many of the items listed above.

'*... the parties involved ...*'. This will include not only the parties to the contract but any outside advisor, subcontractor or third party who may have an interest in the outcome and who should be informed of the adjudication proceedings.

'*... details of where and when it has arisen ...*'. This is somewhat tautological with the brief description of the dispute. However, the circumstances giving rise to the dispute should be given. This should include description of the negotiations which failed to reach a settlement.

'*The nature of the redress sought* ... '. This really is the type of decision that the adjudicator is required to make. For instance, do you want an order that:

(*a*) you should be paid a sum of money;
(*b*) the work is in accordance with the specification;
(*c*) the contractor has not carried out work in accordance with the contract; or
do you want a declaration as to:
(*d*) what the contract means;
(*e*) whether or not work was included in the contract;
(*f*) what rates should be applied to certain variations;
(*g*) whether you are entitled to an extension of time?

These are just examples of some of the remedies you may require.

What you should not ask for is an order which gives an instruction as to what should be constructed. That is not the adjudicator's function, but that of the designer – the architect or engineer.

The redress required should be phrased in terms of contractual obligations not as orders concerning what work should be done.

Finally, the names and addresses of the parties are required so that the appointing body can inform all parties to the dispute of the appointment of the adjudicator, and the adjudicator can write to all parties on his appointment to give instructions on the conduct of the adjudication.

Summary

- Have you described the disputes sufficiently well for it to be clear
  - what the issues are?
  - what experience the adjudicator should have had?
  - who is involved in the dispute?
- Have you described the circumstances giving rise to the dispute?
- Have you stated clearly what decision you want?
- Have you included the names of all parties to the contract?
- Have you sent the Notice of Adjudication to all parties to the contract?

4

The Referring Party: the appointment of the adjudicator and referral of the dispute

The Referring Party has the initiative in appointing the adjudicator.

Who may act as adjudicator?

The answer is almost anyone. However, there are one or two essential qualities required in the selected person. First, it is desirable for the adjudicator to be acceptable to all parties to the dispute. The aim of adjudication is to resolve the dispute and this is most likely to happen if all parties have confidence in the adjudicator.

Second, the adjudicator should have knowledge of both the subject matter of the dispute and the law of the contract. Adjudication is a legal process, and if a party does not like the result he will try to get the decision set aside by the court so that he does not have to comply with it. This will be easiest for the disgruntled party if the adjudicator has either not followed the requirements of the Scheme or has clearly made an error of law.

The Scheme does not say who can be an adjudicator but clearly states who cannot. *Any person requested or selected to act as adjudicator ... shall be a natural person acting in his own capacity. A person requested or selected to act as an adjudicator shall not be an employee of any of the parties to the dispute and shall declare any interest, financial or otherwise in any matter relating to the dispute.*[1]

1 Paragraph 4.

Thus a partnership or company cannot be named adjudicator. However, although the adjudicator should not be an employee of any party, that falls a long way short of requiring him to be independent. For instance, it does not prevent a relative, a person employed by a subsidiary company, or a close friend from acting, although this would be extremely undesirable.

Even though the Scheme states that the adjudicator must declare any such interest it does not say when. It must be before he agrees to be the adjudicator. If the adjudicator subsequently becomes aware of any such interest he should declare it and leave it to the parties to decide whether or not they still want him to continue. If one party objects the other should agree to revoke the adjudicator's authority under paragraph 11(1). The agreement should say how the adjudicator's charges are to be paid. If the adjudicator should have been aware of the clash of interest at the time when he accepted the appointment and failed to declare it then the adjudicator should meet his own costs.

Where the adjudicator finds that he could be seen to have an interest in the outcome of the adjudication he should resign of his own accord.

Making the appointment

Unless the adjudicator is named in the contract then it is the Referring Party who will approach either the appointing authority named in the contract or an adjudicator nominating body (ANB).

An adjudicator nominating body is defined in the Scheme as *a body (not being a natural person and not being a party to the dispute) which holds itself out publicly as a body which will select an adjudicator when requested to do so by a referring party.*[2] This is because it is considered that generally it will be the Referring Party who wishes an early decision and prevents the other party from delaying the adjudication by procrastination.

2 Paragraph 2(3).

The Scheme provides four ways in which the adjudicator can be selected:[3]

(*a*) by agreement between the parties;
(*b*) as specified in the contract;
(*c*) as selected by an appointing body named in the contract; and
(*d*) as selected by an adjudicator nominating body.

This is also the order of priority of choice. It follows a logical sequence — the parties are always free to change what is in the contract, but if they do not agree to do so the contract provisions stand.

Provision is also made in case the person named in the contract is unable or unwilling to act. This makes the wording of the Scheme somewhat clumsy.

The Scheme states:[4]

> *2.(1) Following the giving of a notice of adjudication and subject to any agreement between the parties to the dispute as to who shall act as adjudicator*
>
> (*a*) *the referring party shall request the person (if any) specified in the contract to act as adjudicator, or*
>
> (*b*) *if no person is named in the contract or the person named has already indicated that he is unwilling or unable to act, and the contract provides for a specified nominating body to select a person, the referring party shall request the nominating body named in the contract to select a person to act as adjudicator, or*
>
> (*c*) *where neither paragraph (a) nor (b) above applies, or where the person referred to in (a) has already indicated that he is unwilling or unable to act and (b) does not apply, the referring party shall request an adjudicator nominating body to select a person to act as adjudicator.*

3 Paragraph 2(1).
4 Paragraphs 2(1) & (2).

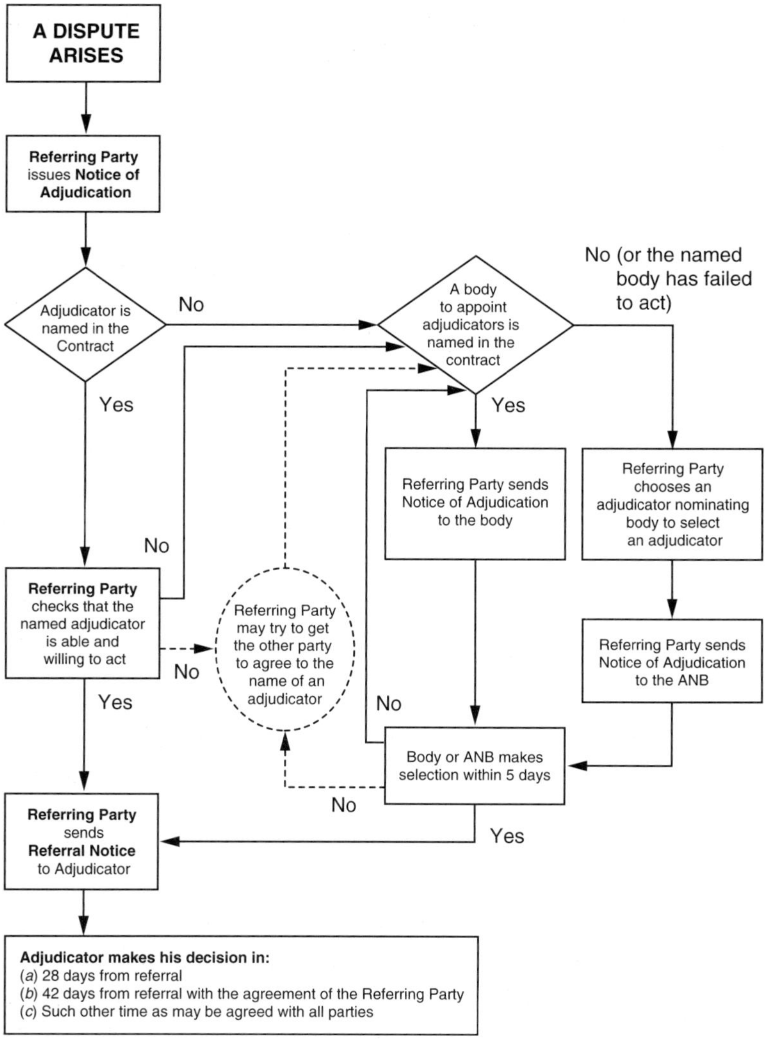

Figure 1. The appointment of the adjudicator

> *(2) A person requested to act as adjudicator in accordance with the provisions of paragraph (1) shall indicate whether or not he is willing to act within two days of receiving the request.*

Adjudicators, however appointed, are given two days within which to agree to act.

When an adjudicator is named in the contract

The Scheme allows for an adjudicator to be named in the contract. It does not matter whether or not the adjudicator is appointed for a non-compliant procedure. He will still be the adjudicator under the Scheme.

Your first action, therefore, after you have decided to refer a dispute to adjudication, is to check whether or not an adjudicator is named in the contract.

If one is, you should contact him to find out whether he is willing and able to undertake the adjudication.

The prospective adjudicator should be sent a copy of the Notice of Adjudication so that he can assess whether or not he is qualified to adjudicate. This should be sent by fax or e-mail to ensure he gets it as quickly as possible.

A problem will arise if you cannot contact the named adjudicator because he is on holiday, or away for some other reason, because the Scheme does not stipulate a time limit.

How long should you allow? The Scheme states that *a person requested to act as adjudicator ... shall indicate whether or not he is willing to act within two days of receiving the request.*[5]

The Act requires that there should be a timetable for the appointment of an adjudicator within seven days of the issue of the Notice of Adjudication.[6] If resort has to be made to an appointing body then that body *must communicate the selection of an adjudicator to the referring party within five days of receiving a request to do so.*[7]

---

5 Paragraph 2(2).
6 s108(2)(b).
7 Paragraph 5(1).

The words 'receiving the request' are not particularly helpful when the adjudicator is named in the contract because it will not always be known when he does, in fact, receive it. If he is away on holiday he might not receive it for two or more weeks if he has not made sufficient arrangements for the receipt of correspondence.

It follows that if the spirit of the Act is to be achieved there are two days in which to get the adjudicator's agreement. However, if he were to become available on day six it would still be possible to refer the dispute to him within the seven days.

It would therefore appear that it is up to the Referring Party to take what action he deems fit, and it is suggested that he would comply with the Scheme if he applied to an adjudicator nominating body if he has not had a response from the adjudicator within 2 days of the request to act.

In any case, all attempts to contact the proposed adjudicator should be recorded and copied to the other party so that there can be no argument later that sufficient effort was not put into trying to obtain the adjudicator's acceptance. Failure to use sufficient effort could lead to the eventual decision not being enforced by the court.

It is debatable whether the Referring Party should ask the prospective adjudicator what his terms and fees for acting will be.

It is always an ideal to have the terms and conditions of any contract settled before performance of that contract starts. However, negotiations can be prolonged and to invite the adjudicator to submit his terms and conditions might enable the other party to delay the start of the adjudication by refusing to agree to the terms or fee.

The adjudicator will, under the common law, be entitled to a reasonable fee for his services and the Scheme provides that he should be paid a reasonable fee.[8] If he charges an amount which the parties consider to be unreasonable then, if the parties refuse to pay it, he will have to apply to the courts for enforcement. The court will decide whether or not the fee requested was reasonable. However,

8 Paragraph 25.

the courts have said they will not substitute their view of the hours that should have been taken by the adjudicator for that of the adjudicator unless there is evidence of bad faith.[9]

If the parties agree between themselves what they consider to be a reasonable fee and pay it, then the adjudicator will have to go to the courts for the difference. This he may be unwilling to do.

There appears to be little risk for the parties if they proceed without an agreement with the adjudicator. The problem may be that the adjudicator will not proceed without an agreement.

If the adjudicator sends his terms for agreement, it may provide a means of procrastination for a reluctant other party.

If speed of decision is important the Referring Party would be advised to accept the terms unilaterally, provided they are acceptable, inform the adjudicator that they are and return the conditions duly signed with the Referral Notice.

If the named adjudicator refuses to act, the Referring Party must approach the body named in the Contract for appointing adjudicators or, if there is not one named or the named adjudicator fails to respond, or another adjudicator is not named in the contract,[10] he can then apply to an ANB.

When an appointing body but not an adjudicator is named in the contract

When no adjudicator is named in the contract, the Referring Party should check if an appointing body is named in the contract.

If so, application must be made to that body for the appointment of an adjudicator.

Most nominating bodies will require a fee before they will act and many have an application form which they require to be completed.

The Referring Party should therefore contact the named body and enquire:

(*a*) whether that body requires applications on a specific application form, and if so where the form may be obtained;

9 *Stubbs Rich Architects* v. *WH Tolley & Sons Ltd* (8 August 2001).

10 Paragraph 6.

(*b*) whether there is a fee, and if so how much;
(*c*) the address to which applications should be made.

All applications to the nominating bodies *shall be accompanied by the Notice of Adjudication.*[11]

If there are any special skills or other characteristics required of the adjudicator which are not clear from the Notice of Adjudication, these special requirements should be made known to the appointing body. These may include:

(*a*) specialist technical knowledge;
(*b*) knowledge of the law;
(*c*) living either close to the site or away from it;
(*d*) names of people who have already been rejected by either party.

Parties should not seek the appointment of a particular individual unless that individual has already been appointed for a connected dispute under a different contract. It may be both economical, time saving and practical for the same person to decide related disputes under two different contracts, since in this way different decisions will not be arrived at by different adjudicators. This is permitted under the Scheme.[12]

When a name is stated, the reason for requesting the named individual should be made clear to the appointing body in the application.

Once the nomination has been made, the Referring Party should contact that person and request confirmation that he is willing to act. If confirmation is not obtained within two days,[13] the referring party may apply to another adjudicator nominating body to select a person to act as adjudicator.

The appointing body may fail to make an appointment within the five days permitted.[14] If that happens *the Referring Party shall request*

11 Paragraph 3.
12 Paragraphs 8(1) & (2) and see below.
13 Paragraph 5(3).
14 Paragraph 5(1).

another adjudicator nominating body to select a person to act as adjudicator.[15]

If the appointing body fails to communicate the selection of an adjudicator to the referring party within five days of being requested to do so, *the referring party may*[16]

(*a*) *agree with the other parties to the dispute to request a specified person to act as adjudicator, or*
(*b*) *request any other adjudicator nominating body to select a person to act as adjudicator.*

The process is similar to that for when no adjudicator or adjudicator nominating body is named in the contract.

When no Adjudicator Nominating Body is specified in the contract

The first action is to select an ANB. The Scheme does not specify any such bodies but permits reference to any body *which holds itself publicly out as a body which will select an adjudicator when requested to do so by a referring party.*[17] A list of some such bodies is to be found in Appendix C.

Once a body has been selected, the Referring Party should first ring up the appointing body and ask:

(*a*) Have you an application form for the appointment of adjudicators? There may be different forms according to whether the adjudicator is required for a contractual adjudication or for an adjudication under the Scheme.
(*b*) Can the application form be:
  (*i*) sent to the Referring Party by fax; or e-mail, or
  (*ii*) downloaded from a website? If so, what is the address of the website?
(*c*) What is the amount of the fee?

15 Paragraph 2(1)(c).
16 Paragraphs 5(1) & (2).
17 Paragraph 2(3).

Once this information has been obtained and the Referring Party has acquired the appropriate application form, it should be completed and returned to the ANB together with:

(*a*) the fee;
(*b*) a copy of the Notice of Adjudication.[18]

Before making a nomination, most nominating bodies will check that the person selected is willing and able to act and that the person considers that he has the necessary skill and expertise.

Again, once the selection has been made, the Referring Party should contact that person and obtain confirmation that he is willing to act. Such confirmation must be given within two days of being requested.[19] If the adjudicator nominating body does not communicate the selection within five days of being requested to do so,[20] the Referring Party may then request any other adjudicator nominating body to select a person to act as adjudicator.[21]

And so on until an adjudicator is appointed.

What the Scheme fails to address is what happens if an adjudicator nominating body communicates a selection after the five days and before the Referring Party has made a new request to another body. It is suggested that the selected person should be sent the Referral Notice and if the other party does not make an immediate objection the selected person would become the adjudicator for the dispute. But it is for the Referring Party to decide either to persist with the selected person or to request another selection.

When such a problem occurs there is a potential inconsistency in the Scheme. It requires that the *referring party shall, not later than 7 days from the date of the notice of adjudication, refer the dispute in writing (the referral notice) to the adjudicator.*[22]

18 Paragraph 3.
19 Paragraphs 2(2), 5(3) & 6(2).
20 Paragraphs 5(1).
21 Paragraph 5(2)(b).
22 Paragraph 7(1).

This will be well-nigh impossible if an adjudicator named in the contract is unable or unwilling to act, or indeed if an ANB does not select an adjudicator within the five days allowed.

The Scheme fails to address this problem.

It is suggested that, unless another application has been made to an ANB, the procedure should continue and the Referral Notice be delivered late. In this way neither party would be prejudiced since the Referring Party could always apply to another ANB requesting the same person to be appointed on the grounds that the only reason for the failure to appoint was due to the time constraint. It will also save an unnecessary appointment fee.

Summary

The Referring Party should:

- Check whether an adjudicator is named in the contract.
  - If so, send a copy of the Notice of Adjudication to that person and request confirmation of ability and willingness to act.
- If not, check whether an adjudicator appointing body is named in the contract.
  - If so, apply to that body for an application form and complete it.
- If not, select an ANB (from Appendix C) and apply for an application form.
- Applications should be accompanied by:
  - the Notice of Adjudication;
  - the appropriate Fee.

Actions after the appointment is confirmed

Once the adjudicator has confirmed that he is willing to undertake the adjudication, the next step is for the Referring Party to send to the Adjudicator the Referral Notice.

The Scheme states:

Where an adjudicator has been selected ... the referring party shall, not later than seven days from the date of the notice of adjudication, refer the dispute in writing (the 'referral notice') to the adjudicator.[23]

The Referral Notice

The Referral Notice starts the adjudication time clock. However, time starts from the date of the notice of adjudication.[24] It is not particularly clear when this is. The date must be the date that the Notice of Adjudication is served on the Responding Party, which is not necessarily the same as the date on the Notice itself. The Responding Party should therefore date stamp the Notice with the date when it was received and acknowledge receipt of it stating that date.

Referral occurs when the adjudicator receives the Referral Notice.

'Referral Notice' is a misnomer. The Referral Notice is intended to be not just a notice but full details of the arguments and supporting information concerning the claim being made by the Referring Party. It is in effect his statement of case.

A referral notice shall be accompanied by copies of, or relevant extracts from, the construction contract and such other documents as the referring party intends to rely on.[25]

The clear intention is that the Referral Notice should be accompanied by a full statement of the Referring Party's case. Full information of the claim is required so that both the adjudicator and the Responding Party have the full 28 days to make a considered reply and for the adjudicator to reach his decision.

It has to remembered by the Referring Party that the Responding Party must be given a reasonable opportunity of presenting his case

23 Paragraph 7(1).
24 Paragraph 7(1).
25 Paragraph 7(2).

otherwise the adjudicator's decision may not be enforced by the courts for failure to follow the rules of natural justice.[26]

You should therefore, if you are the Referring Party, prepare in advance of the Notice of Adjudication a comprehensive statement of your claim so that this can be sent to the adjudicator immediately after you serve the Notice of Adjudication. The Referral Notice should consist of the following:

(*a*) Details of the contract:
- (*i*) the title of the contract and what it is for;
- (*ii*) the parties to it;
- (*iii*) when it was formed and its value;
- (*iv*) the standard conditions of contract (if any) upon which it is based, or a statement that one of the parties drafted them.

(*b*) The background to the Claim:
- (*i*) how it arose;
- (*ii*) steps taken to resolve it;

(*c*) The relevant facts concerning the events leading to the claim with documentary support:
- (*i*) signed statements of evidence from the people directly involved;
- (*ii*) photographs;
- (*iii*) site records;
- (*iv*) relevant extracts from the contract: clauses, specification, bills of quantities, site information, etc.

(*d*) The contractual arguments as to why you are entitled to what you are asking.

(*e*) Comments on the other party's position as you understand it.

(*f*) What you want the adjudicator to include in his decision.

Care should be taken to see that only that which is relevant to the dispute is sent to the adjudicator. He does not want the whole of the contract, the whole of the specification and all the drawings. He just requires those extracts which are relied upon by each party.

26 *Russell* v. *Duke of Norfolk* (1949) m1 All ER 109.

He will not normally read the whole contract if it is sent to him. He will just read those parts referred to in the statements.

If he does read the whole contract he will charge for the time taken reading it and the parties will incur unnecessary expense.

You should include all the information that you consider relevant because you may not have a further opportunity. The Scheme gives no automatic right of reply to the other party's case and the adjudicator may proceed to a decision immediately after receipt of comments from the Responding Party.

If you think the other party will raise further contractual matters which you do not wish to address, you should warn the adjudicator that you will want an opportunity to reply and ask for it in the Referral Notice.

A badly prepared case gives the adjudicator an unfavourable impression, and may adversely prejudice the decision. Time and care in preparation of the case is time and money well spent.

Copies of the Referral Notice and its supporting documents should be sent to the adjudicator and *the referring party shall, at the same time ... send copies of those documents to every other party to the dispute.*[27]

A point to note is that correspondence between the parties and the contract administrator written during the execution of the work often takes important facts for granted because the writer and recipient are familiar with the facts. The adjudicator is not and many letters may need additional explanation.

It is then up to the Adjudicator to decide how to conduct the adjudication.

Conduct of the adjudication

The Scheme gives no guidance as to how the adjudication should be conducted. It does not even provide for the submission of a defence by the Responding Party. Clearly, this has to be allowed but at what stage the defence should be delivered and in what form is not stipulated. It is up to the Adjudicator.

27 Paragraph 7(3).

The Scheme gives the adjudicator wide choice as to how to proceed. This is detailed and described in Chapter 6. It will be necessary for the adjudicator to decide immediately upon receipt of the Referral Notice how he intends to proceed, and what further information he requires.

Therefore, when writing the Referral Notice, the Referring Party should consider any procedures that he would like the adjudicator to adopt and suggest these procedures in the Notice.

Thought will always have to be given to the length of time that the other party can have in which to reply. If the Referral Notice is very long and consists of a considerable amount of newly drafted evidence which the other side has not seen, the Responding Party will require a greater time to deal with it than if it is merely a restatement of information already submitted.

The Referring Party should not make unreasonable demands upon the other side as this is likely to boomerang. If the Receiving Party or the adjudicator are allowed insufficient time to undertake their work the adjudicator will take this into account when reaching his decision.

Therefore, propose a reasonable timetable.

Other procedural matters should also be considered. Proposals for the procedure may have been considered in the run-up to the adjudication and either no agreement reached or the proposals may have been rejected by the other side. There is no reason why such matters should not be put to the adjudicator for a decision.

For instance, if it is considered desirable that further tests should be made and these have so far been refused by the other side, the adjudicator can be asked to arrange for the necessary tests to be carried out. Similarly, if it is thought that advice — legal or technical — would be of assistance, it would be sensible for the adjudicator, as a neutral person, to obtain that advice on behalf of both parties, thereby saving the costs of a second person.

Combining disputes

The adjudicator can consider only those disputes that have been referred to adjudication in the Notice of Adjudication. Negotiations often involve many issues. When one party decides that it is not making progress on one or more of the issues, it may refer those issues to adjudication. The other party may then wish to refer other issues which he thinks are not progressing to adjudication at the same time. He must then issue another Notice of Adjudication detailing the issues that he wishes to refer.

However, unless all the disputes are in one Notice of Adjudication the adjudicator cannot necessarily deal with them all in the same adjudication. Indeed, it is quite possible that different adjudicators are appointed for the different issues. It would normally be more efficient and economical for both sets of issues to be dealt with by the same adjudicator at the same time.

The Scheme permits this if all parties to the disputes agree: *The adjudicator may, with the consent of all the parties to those disputes, adjudicate at the same time on one or more disputes under the same contract.*[28]

One other very important matter to consider is what is called 'joinder'. Often a dispute under one contract is identical to a dispute under another. For instance, if a subcontractor performs some work which the client thinks is not in accordance with the specification, two identical disputes will probably arise — one under the main contract and one under the subcontract.

If these two disputes are referred to two different adjudicators, two different solutions will probably result, and you will certainly pay for the same arguments to be considered twice. There is therefore every incentive for the two disputes to be considered together by the same adjudicator.

The Scheme provides for this also: *The adjudicator may, with the consent of all the parties to those disputes, adjudicate at the same time on*

28 Paragraph 8(1).

related disputes under different contracts, whether or not one or more of these parties is a party to those disputes.[29]

This is rather curiously worded but the essential point is that not all parties have to be parties to all the disputes.

The combining of more than one dispute into a single adjudication produces two additional problems.

(*a*) Because the Notices of Adjudication for the disputes will not have been issued on the same date there is a problem of the date by which the decision must be made.
(*b*) Separate adjudicators may have been appointed for each dispute and because of the combining of the disputes one or more adjudicators is made redundant.

Again, the Scheme deals with these two problems.

It allows all the parties to the disputes to agree to extend the times. *All the parties in paragraphs (1) and (2)* (covering combining disputes and joinder) *respectively may agree to extend the period within which the adjudicator may reach a decision in relation to all or any of these disputes.*[30]

Again, the essential point is that all the parties to all the disputes must agree to the timetable. Where it is the Referring Party who is the party involved in two or more different disputes under either the same or different contracts he should consider whether or not he wishes the disputes to be combined and, if he does, advise the adjudicator and the other parties in the Referral Notice. Where a second contract is involved, it will probably mean sending additional copies of the Referral Notice to the other parties to that second contract.

The parties can always agree, at any time, to revoke the appointment of an adjudicator[31] and the adjudicator is then entitled to be paid such reasonable amount as he may determine. However,

29 Paragraph 8(2).
30 Paragraph 8(3).
31 Paragraph 11(1).

in respect of the adjudicator ceasing to act because of the joining of two or more disputes, for the avoidance of doubt, the Scheme deals with them in identical wording to normal revocation but under a different paragraph: *Where an adjudicator ceases to act because a dispute is to be adjudicated on by another person in terms of this paragraph, that adjudicator's fees and expenses shall be determined in accordance with paragraph 25.*[32]

Any agreements to consolidate or join disputes must include an agreement as to the date for delivery of the decision. It would also be prudent to agree a timetable and a procedure.

Negotiations for any such agreement must be concluded quickly because if agreement proves impossible the original timetable for the adjudication stands unless the parties agree to extensions of time. If no extension of time is agreed by the Referring Party the Responding Party will suffer because it will be his time for reply which will be curtailed.

Summary

- If an adjudicator is not named in the contract:
  - Is there a person named in the contract to select an adjudicator?
  - If not, the Referring Party chooses an ANB and applies for a selection.
  - Nominating bodies must make a selection within five days.
- The Responding Party cannot object to a selection.
- Prospective adjudicators must confirm within two days that they are able and willing to accept the appointment.
- The Referral Notice is the Referring Party's full statement of his case.
- The Referral Notice should contain:
  - All the information he wishes to bring to the attention of the adjudicator;

32 Paragraph 8(4). It should be noted that paragraph 25 is identical in wording to the second and third sentences of 11(1).

- comments on the anticipated defence of the Responding Party;
- proposals on the timetable for the adjudication;
- proposals for any joining of disputes;
- a copy of the Notice of Adjudication;
- a request for reasons (but only if reasons are required).

5

The Responding Party: receipt of the Referral Notice

The Responding Party may or may not wish to have a speedy adjudication of the matters in dispute. However, the possibilities for employing delaying tactics are very limited.

The Responding Party should therefore begin to prepare his case immediately the Notice of Adjudication arrives because time will be very limited once the adjudicator is appointed. Normally, before the Notice of Adjudication is served the parties will have discussed fully the issues involved and the Responding Party will be reasonably well informed of the issues when the Notice arrives.

Actions on receipt of Notice of Adjudication

On receipt of the Notice of Adjudication the Responding Party should check that:

(*a*) He recognises the issues in dispute, and understands what it is that the Responding Party requires, and that

(*b*) the Notice of Adjudication states all the information required by the Scheme. That is:[1]

   (*i*) *The nature and a brief description of the dispute and of the parties involved,*

   (*ii*) *Details of where and when it has arisen,*

1 Paragraph 1(3)(a)–(d).

(*iii*) *The nature of the redress which is sought, and*
(*iv*) *The names and addresses of the parties to the contract (including, where appropriate, the addresses which the parties have specified for giving notices).*

Since these items (*i*)–(*iv*) are mandatory requirements for the Notice of Adjudication under the Scheme, any omission may invalidate the Notice. The Responding Party must then decide whether to accept the Notice or ask for it to be reissued correctly.

Any error should be brought to the attention of the Referring Party so that it can be clarified or remedied. However, it is suggested that it is unlikely that a court would set aside an adjudicator's decision because the Notice of Adjudication was deficient in some regard, particularly if the Responding Party subsequently responds to it.

It is particularly important that the nature of the dispute is clearly understood so that the statement expressing the views of the Responding Party can be written. If it is not clear, the Referring Party should be asked to clarify it.

The other matters required in the Notice of Adjudication are less important, since they do not affect the preparation of these documents but are intended to inform a prospective adjudicator or a nominating body (ANB), if one is to make a selection, of the nature of the dispute. This will enable the prospective adjudicator to decide whether he is competent to decide the issues, and an ANB to have sufficient knowledge to be able to select a suitably qualified person.

There is little point in challenging the validity of the issues at this stage, if the Responding Party wishes to delay the appointment of the adjudicator, as most ANBs will not investigate such allegations but will leave it to the appointed adjudicator to investigate.

There is little chance that the appointment of an adjudicator can be prevented.

When the Responding Party is satisfied that he understands the matters in dispute he should start to prepare his response so that when the Referral Notice arrives most of the basic information will have been gathered.

Combining disputes

The adjudicator can consider only those disputes referred to adjudication in the Notice of Adjudication.

The Responding Party may on receipt of the Notice of Adjudication suggest that other issues which are the subject of current negotiations should also be referred to adjudication if they have not been selected by the Referring Party.

The Responding Party should then issue another Notice of Adjudication for those other disputes if he wishes the matters to be considered at the same time. The reason is that disputes not mentioned in the Notice of Adjudication cannot be considered by the adjudicator unless all parties to the dispute agree. The longer it takes for a further Notice of Adjudication to be issued the less likely it is that the first Referring Party will be willing to agree to the disputes being considered together.

The Scheme allows different disputes under one contract to be combined with the agreement of all the parties.[2] This is fully explained in Chapter 4.

However, unless all the disputes are in one Notice of Adjudication the adjudicator cannot necessarily deal with them all in the same adjudication. Indeed it is quite possible that different adjudicators are appointed for the different issues. It would normally be more efficient and economical for both sets of issues to be dealt with by the same adjudicator at the same time.

The Scheme permits the parties to combine their disputes if all parties to the disputes agree: *The adjudicator may, with the consent of all the parties to those disputes, adjudicate at the same time on one or more disputes under the same contract.*[3]

Appointment of the Adjudicator

Unless the adjudicator has already been agreed either:

(*a*) by naming him in the contract, or

2 Paragraph 8(1).
3 Paragraph 8(1).

(*b*) by agreement between the parties after the contract has been entered into,

the Responding Party has little influence over the appointment of the adjudicator. This is because without such agreement it is up to the Referring Party to arrange the appointment and he may:[4]

(*a*) if he wishes, ask the Responding Party to agree to an adjudicator,
(*b*) apply to the person named in the contract for appointing an adjudicator to make an appointment, or
(*c*) if no body is named, ask an ANB (see Appendix C) to appoint an adjudicator.

An ANB is defined in the Scheme as *a body (not being a natural person and not being a party to the dispute) which holds itself out publicly as a body which will select an adjudicator when requested to do so by a referring party.*[5]

It is probable, therefore, unless the adjudicator is named in the contract, that the Responding Party will learn the name of the adjudicator only when the Referral Notice arrives.

Actions when the Referral Notice arrives

When the Referral Notice arrives the Responding Party should:

(*a*) Check that the contract is one to which the Act applies. For this, reference must be made to the Act ss104–107. If the contract is in relation to construction works then probably the only check necessary is to see whether or not the contract falls within the exclusions in s105 (see Appendix A for details).
(*b*) Check whether or not there is an adjudication clause in the contract.
   (*i*) If there is, check whether the clause complies with the eight compliance points[6]

4 Paragraph 2(1)(a)–(c)
5 Paragraph 2(3)
6 Chapter 1 p4.

(*ii*) If it does, the contract adjudication clause will apply and not the Scheme. The remainder of this book will then not apply, although much of it will still be useful advice. There will probably be an Adjudication Procedure referred to in the clause and that Procedure will then have to be followed.

(*iii*) If the clause does not comply and if either party so wishes, the adjudication may be carried out in accordance with the Scheme.

It should be noted that even if a contract clause does not include the eight compliance points, this does not render the clause void. It is still a valid adjudication provision but the Act gives the parties a right to adjudication in accordance with the Scheme. So the choice is with either party to decide whether or not to invoke that right. It is, of course, possible to have two adjudications concurrently — one in accordance with the contract clause and one in accordance with the Scheme.

That would be absurd and the parties should agree to proceed under the Scheme because, if it proved necessary to obtain an enforcement order from the court, it is likely that the court would give precedence to the adjudication under the Scheme.

(*iv*) If there is no adjudication provision in the Contract the Scheme will apply.

(*c*) Check that the correct procedure for appointing the adjudicator has been followed:

(*i*) Is there an agreed adjudicator either named in the Contract or agreed afterwards?

(*ii*) If not, is there a specified nominating body named in the Contract?

If so, has the appointment been made by that body?

(*iii*) If not, has the Responding Party selected an ANB?

(*iv*) If so, has the appointment been correctly made?

(*v*) In any case where the proper procedure has not been followed, the Responding Party must decide whether or not to accept the appointment.

(*vi*) If he decides against, the Responding Party should write to the appointed person informing him that he has been wrongly appointed, and explain why. Photocopies of the contract provisions which demonstrate the wrong procedure has been followed should be included.

However, if you are prepared to accept the appointed adjudicator, write to him and the Responding Party to confirm that you are accepting the appointment in spite of the wrong procedure being adopted.

It will be up to the appointed person to decide whether or not to continue with the adjudication. If he does the Responding Party must decide whether or not to continue with the adjudication. If he wishes to do so, he should tell the adjudicator that he is reserving his position with regard to the validity of the appointment before taking any step in the adjudication. If he fails to do this he will probably lose the right to complain later. That is certainly the position under the Arbitration Act 1996.[7]

(*d*) If the appointment has been correctly made, check that the appointed person has no conflict of interest with any of the parties, or that there are no grounds for considering that he may be prejudiced in some way concerning the subject matter of the disputes. Any objection to the appointment must be made as soon as the appointment is made, but will not invalidate the appointment.[8]

(*e*) Check that the issues in the Referral Notice are the same as those in the Notice of Adjudication.

If there are different or additional issues the adjudicator has no authority to deal with them unless:

(*i*) the Responding Party and the adjudicator agree that they can be considered,[9] or

7 s73 Arbitration Act 1996.

8 Paragraph 10.

9 Paragraph 9(1).

(*ii*) they are matters that the adjudicator *considers are necessarily connected with the matters in dispute*,[10] or
(*iii*) an additional Notice of Adjudication is issued and the same adjudicator appointed for the new issues.

Since matters under (*i*) can be decided only with the agreement of the parties, the Responding Party, in effect, can decide whether or not they should be included in the adjudication.

There is no need to take separate action at this stage as all these matters can be addressed in the reply statement.

The Statement in Response

The Scheme does not refer expressly to a Statement from the Responding Party. The only reference to the Responding Party being entitled to submit a statement is that *the adjudicator shall consider any relevant information submitted to him by any of the parties to the dispute and shall make available to them any information to be taken into account in reaching his decision.*[11]

The Responding Party should therefore make clear to the adjudicator that he is intending to make a reply to the Referral Notice and say when that reply is likely to be sent.

The Responding Party should also consider what information it will be necessary to give to the adjudicator in order for him to reach a decision. This may involve:

(*a*) obtaining written statements from people who are, or were, engaged on the construction contract;
(*b*) obtaining documents from storage or third parties such as consultants;
(*c*) arranging for additional tests or inspections;
(*d*) consulting advisers such as lawyers, consultants, or users;
(*e*) holding a meeting with the adjudicator whether with or without the other party being present.

10 Paragraph 21.
11 Paragraph 17

Since it is the adjudicator's duty to set the timetable and procedure, he will require assistance from both parties in deciding the best procedure for the particular dispute. He will need to make this decision right at the outset. The Responding Party should therefore contact the adjudicator with his comments as soon as possible after receipt of the Referral Notice to discuss the procedure to be adopted.

There is no reason why this should not be by telephone. If the adjudicator considers that he will require more time than the 28 days he will wish to know that right from the outset. He can then arrange firstly with the Referring Party for a 14 day extension[12] or if a longer one is necessary make arrangements with all parties.[13]

At every stage of the preparation of the Response Statement the requirements of the adjudicator should be borne in mind.

The adjudicator has to compare and decide between the cases prepared by the parties, and that task is very difficult if the Response Statement does not address every point made in the Referral Notice.

The points on decision-making listed in Chapter 1 should be borne in mind, particularly the principle that 'what is not denied is accepted'. Even if you disagree entirely, for instance, with the evaluation of a claim for additional payment and the basis upon which it has been calculated, you should nevertheless examine the basis of all the figures quoted in case the adjudicator decides against you. If you have not criticised the figures he would have no alternative but to accept the figures as presented.

Go through the Referral Notice paragraph by paragraph and accept, deny or comment on each.

Where you have an alternative argument always give the reason under the contract (referring to the clause by number) that supports your case. Generalisations and reliance solely on reasonableness are of very limited value. The adjudicator has to decide the dispute strictly in accordance with the contractual entitlements of the parties. If you do not help him identify why you have an entitlement, you are likely to lose.

12 Paragraph 19(1)(b).
13 Paragraph 19(1)(c).

Therefore, examine the Referral Notice first and then present any alternative arguments that you may have.

The more documents you can produce to support your case the better, but do take care to see that only relevant documents are given to the adjudicator. He will have limited time in which to reach a decision and he will take less notice of your submission if much of it is irrelevant.

If when you have finished the Response Statement you consider that it would be helpful to have a private meeting with the adjudicator to explain your case you should ask for one.

When the statement is complete, check it for typographical and other errors and submit it to the adjudicator; at the same time, send a copy of all that you send to the adjudicator to all the other parties to the dispute.

Summary

- On receipt of the Notice of Adjudication:
  - Check whether you have an adjudication clause in the contract.
  - If so, check whether it complies.
  - If not, check that the Act applies to your contract.
  - If so, check that the Notice of Adjudication is valid.
- When the Referral Notice arrives:
  - Check its contents comply with the Notice of Adjudication.
  - Answer each point made in it with a denial, acceptance or comment.
  - Keep your alternative arguments separate from the responses to the Referral Notice.
  - Criticise all figures including those where you consider there is no entitlement.
  - Always aim to assist the adjudicator.
  - Send a copy of all you send to the adjudicator to all other parties.
  - Request reasons (but only if reasons are required and have not been requested by the Referring Party).

6

The procedure — determination by the adjudicator

The Scheme is silent on the procedure to be adopted, leaving it to the adjudicator and the parties to determine.

The only requirement is for the Responding Party to provide the arguments and information concerning the dispute with the Referral Notice.[1]

There is no requirement for the Responding Party to submit anything.

The adjudicator shall:

(*a*) *act impartially in carrying out his duties and shall do so in accordance with any relevant terms of the contract and shall reach his decision in accordance with the applicable law in relation to the contract; and*

(*b*) *avoid incurring unnecessary expense.*[2]

but may not take into account in reaching his decision any information that is not made available to all parties to the dispute.[3]

The meaning of the second phrase of paragraph (a) (above) from the Scheme, namely *and shall do so in accordance with any relevant terms of the contract*, is ambiguous. It is not clear whether the words require the adjudicator to comply with any procedure for

1 Paragraph 7(2).
2 Paragraph 12.
3 Paragraph 17.

adjudication that may be in the contract or whether they refer to the basis upon which he shall reach his decision.

It would be surprising if the words were intended to refer to any adjudication procedure, since the Scheme will apply only where the contract provision for adjudication does not comply with the requirements of s108 of the Act.

This means that the Referral Notice, and the evidence contained in it must address the contractual rights of the parties and not be based upon common sense.

It is therefore more probable that they were intended to refer to the terms of the contract relevant to the dispute, and read in conjunction with the last words would seem to imply that the adjudicator must decide the dispute in accordance with the legal rights of the parties. This is made clear in the Scottish version of the Scheme.

Because the adjudicator, in making his decision, can only take into account information known to all parties, the parties should ensure that all correspondence with the adjudicator is communicated to all other parties.

A check that all parties have received all information sent to the adjudicator is normally achieved by the adjudicator acknowledging receipt of all documents received by him to all parties.

If the adjudicator fails to acknowledge any correspondence the sender should enquire whether it has been received.

From the start all parties must be conscious of the time available for both the submission of information and the reaching of the decision by the adjudicator. These are:[4]

28 days from receipt by the adjudicator of the Referral Notice, or
42 days if the Referring Party agrees, or
any period exceeding 28 days with the agreement of all parties.

The Receiving Party must accept and be prepared from the outset for the Responding Party not to agree to an extension of these

4 Paragraph 19(1).

periods, and should start preparation of his statement immediately the Notice of Adjudication arrives.

Both parties should be mindful of the time periods and consider the timetable for the adjudication. The Referring Party would be unwise to insist upon the timetable that would not give his opponent adequate time to respond.

If a party has inadequate time to prepare his case, this is a fact that the adjudicator may take into account when reaching his decision. Adjudicators do not view favourably the tactics that are aimed at disadvantaging the other party.

The range of disputes and their individual requirements is infinite. Construction disputes range from single issue disputes on, for instance, a rate for an item of work, to disputes of a complex technical nature. Issues may be a matter of opinion upon which the adjudicator is already well versed or they may require additional tests or investigations to be made. Some problems may require oral evidence, for instance, what was said at a meeting, and others may be dealt with adequately on documents only. The adjudicator may want to meet the witnesses to form an opinion on their reliability.

The appropriate procedure for one may not be appropriate for another. It is for the adjudicator to decide.

It is clear from the Scheme that the adjudicator is not expected to rely only upon the evidence and submissions of the parties. The Act requires that he shall be enabled to take the initiative in ascertaining the facts and the law.[5]

The Scheme repeats the words from the Act. *The adjudicator may take the initiative in ascertaining the facts and the law necessary to determine the dispute, and shall decide on the procedure to be followed in the adjudication. In particular he may ...*[6]

The intention must be to put the adjudicator under an obligation to satisfy himself on the essential features of each party's case before reaching his decision. Any person who has been involved in dispute resolution will confirm that, normally, cases develop as they proceed

5 s108(2)(f).

6 Paragraph 13 opening sentence.

and often the issues debated at the end are considerably different from those discussed at the beginning. This is because parties do not necessarily put forward alternative arguments at the beginning for fear of detracting from the force of the arguments they are putting forward.

However, it must be remembered that it is the dispute as it existed at the time of the issue of the Notice of Adjudication that is put to the adjudicator, not some entirely new dispute.

The Scheme requires the adjudicator to reach his decision in accordance with the applicable Law.[7] It is to be a decision based upon the legal rights of the parties.

Such a decision requires the adjudicator to establish the relevant facts on the basis of the balance of probabilities and to establish the relevant law. The established facts are then applied to the relevant law to reach the decision.

It is probable that the parties will in their submissions provide insufficient information for the adjudicator to be able to establish either the relevant facts or the law.

The Scheme, therefore, provides the adjudicator with express powers to assist him to ascertain the facts and the law.

However, the adjudicator will take the initiative only in ascertaining the facts and the law to the extent that is necessary on the statements that he has received. Parties should not expect the adjudicator to make their cases out for them. For instance, the adjudicator will assume the parties have drawn to his attention all the clauses in both the contract and the specification that are relevant to the dispute. He will not examine either of them just to satisfy himself that the parties have not missed an essential point. Nor will he do an exhaustive enquiry to ensure that he has been told all the relevant facts.

... and shall decide the procedure to be followed in the adjudication.

This sentence clearly makes the adjudicator master of the procedure to be adopted.

7 Paragraph 13.

The first task of the adjudicator on receiving the Referral Notice is to read it carefully and to decide what procedures should be adopted. Clearly it is in the best interests of the parties that the procedure should be acceptable to them. The adjudicator should therefore ascertain the views of the parties as to how they think it would be best to proceed.

He may do this by phone or fax, but it is essential that the procedure should be decided as quickly as possible in view of the time limits.

All parties and the adjudicator should consider from the outset how best to present their case, and what it is necessary for the adjudicator to know. If either party considers:

(*a*) further tests are necessary;
(*b*) a site visit will be desirable — perhaps immediately;
(*c*) the subject matter will be presented most effectively orally;
(*d*) witnesses, samples, plant or equipment will only be available on specific days;
(*e*) expert opinion evidence will be of assistance,

that party should say so to the adjudicator as early in the procedure as possible.

This applies particularly to the Responding Party who may want two or three weeks to prepare his case because it will be too late for the adjudicator to make the necessary arrangements if such information is given to him only when the Responding Party submits his response.

In order to give the adjudicator the ability to perform his duties the Scheme lists a wide range of powers,[8] and although these are intended to be decided by the adjudicator there is no reason why the Referring Party should not make suggestions to the adjudicator in the Referral Notice. Similarly the Responding Party may make suggestions on procedure in his response, or earlier if urgency makes this appropriate.

8 Paragraph 14.

The Scheme permits the adjudicator to take various courses. He can:[9]

13(*a*) *request any party to the contract to supply him with such documents as he may reasonably require including, if he so directs, any written statement from any party to the contract supporting or supplementing the referral notice and any other documents given under paragraph 7 (2).*

Again, this is curiously worded because it would seem to imply that the information will come from the Referring Party, whereas, in fact, it is more likely to come from the Responding Party.

The parties should consider what documents are relevant to the dispute and may be helpful to their case, which they do not have and which the other party possesses. Such documents could be correspondence with third parties, copies of orders, site records, or the minutes of internal meetings. All these documents would probably have to be revealed if the matter was subsequently taken to a court or arbitration.[10] Little will be gained by withholding requested documents, and considerable costs might be incurred if the dispute is referred subsequently to a court or arbitrator.

The parties are required to comply with any request by the adjudicator. The Scheme states: *The parties shall comply with any request or direction of the adjudicator in relation to the adjudication.*[11]

Failure to comply with any request or to produce any document or statement requested by the adjudicator can result in substantial disadvantage. Under the Scheme if a party fails to comply with a direction or instruction of the adjudicator, he may take that into account.

The Scheme states: *If without showing sufficient cause, a party fails to comply with any request, direction or timetable of the adjudicator made*

9 Paragraph 13 (a)–(h).

10 This process is known as disclosure, and courts and arbitrators have power to order the disclosure of any relevant document which might open an avenue of enquiry relevant to the matters in dispute.

11 Paragraph 14.

in accordance with his powers, fails to produce any document or written statement requested by the adjudicator, or in any other way fails to comply with a requirement under these provisions relating to the adjudication the adjudicator may:[12]

(*a*) *continue the adjudication in the absence of that party or of the document or written statement requested,*
(*b*) *draw such inferences from the failure to comply as the circumstances may, in the adjudicator's opinion, be justified, and*
(*c*) *make a decision on the basis of the information before him attaching such weight as he thinks fit to any evidence submitted to him outside any period he may have requested or directed.*

Failure by a party to provide a document requested can be serious because the Scheme entitles the adjudicator to speculate as to the content of a document not submitted. For instance, when a sum of money is claimed to have been paid to a third party and a request is made for a copy of the receipt and is not forthcoming, the adjudicator may assume that the sum has not been paid.

The adjudicator's decision is binding until such time as the dispute may be finally determined by legal proceedings, arbitration, or agreement.[13]

The consequence may be that a party will have to embark on a legal process to get a decision of the adjudicator changed. This will, in any case, result in considerable unrecoverable costs. Parties who win recover generally only about 60% of their costs.

Failure to comply with an instruction of an adjudicator can therefore have far-reaching effects.

These sanctions also apply to non-compliance with any direction or instruction of the adjudicator not just for failure to produce a document.

13(b) decide the language or languages to be used in the adjudication and whether a translation is required and by whom.

12 Paragraph 15.
13 s108(3) of the Act.

This is a power given to the adjudicator which the parties cannot overrule. It will generally not arise except where the preferred language of the parties is not English.

In such cases the parties must agree an adjudicator who is fluent in the preferred language because it is by no means possible that the appointing authority, or ANB will be willing or able to appoint a person fluent in that language.

13(c) *meet and question any of the parties to the contract and their representatives.*

Meetings can be very useful in clarifying the issues that the adjudicator has to decide. Often the initial statement of case (the Referral Notice) does not set out the problem in a manner that is capable of decision. The adjudicator must therefore agree with the parties the exact questions that he is expected to answer. It is often easier and quicker to get agreement on the issues at a meeting than by correspondence.

If the adjudicator is unclear what he has to decide he should call a meeting as early as possible to reach agreement with the parties. Such agreement is essential when the actual issues differ from those set out in the Notice of Adjudication because failure to obtain agreement could lead to argument later should one party not wish to implement the adjudicator's decision.

A second situation for calling a meeting is when there are a number of complex issues.

Complex issues are often better explained orally rather than relying totally upon documents. The Scheme permits the parties to be represented by whomsoever they wish:[14]

16(1) *Subject to any agreement between the parties to the contrary, and to the terms of paragraph (2) below, any party to the dispute may be assisted by, or represented by, such advisers or representatives (whether legally qualified or not) as he considers appropriate.*

14 Paragraph 16(1) & (2).

(2) *Where the adjudicator is considering oral evidence or representations, a party to the dispute may not be represented by more than one person, unless the adjudicator gives directions to the contrary.*

The Scheme appears to permit the adjudicator to meet the parties separately and this is often the best way of getting a full understanding of a party's case. However, such a procedure is fraught with danger. It is an essential feature of natural justice that each party knows the case it has to answer. The courts have failed to enforce three decisions on the grounds of breach of natural justice. In two the adjudicator obtained or might have obtained information unknown to the other party because in one case he tried to mediate the dispute and when the mediation failed reverted to being an adjudicator.[15] In another case the adjudicator entered into extensive telephone conversations with the representatives of one of the parties and the court held that there was a serious risk of bias and failure to make the content of the telephone conversations available to the other party constituted a breach of natural justice.[16]

If the adjudicator does see the parties separately the adjudicator should ensure that:

- he gets the agreement of all parties in writing to his seeing them separately;
- he provides the same opportunity to all parties;
- he makes sure that all parties are aware of any information gleaned in such sessions which may influence his decision, and that they are given an opportunity to comment on it;
- he should inform them and get their acceptance of the fact that he might overlook some details and fail to inform them of them;
- he should also take careful notes of all that transpires.

15 *Glencot Development and Design Co Limited* v. *Ben Barrett & Son (Contractors) Limited.*
16 *Discain Project Services Limited* v. *Opecprime Developments* [2000] BLR 402.

Provided that they are approached with care, such meetings can be beneficial.

At such meetings it will be the adjudicator that asks the questions because the purpose of the meeting is for him to ascertain the facts and the law. Any witnesses or other persons from whom the adjudicator requests information will be questioned by the adjudicator and not examined and cross-examined in the manner prevailing in court or arbitration.

Adjudication is an enquiry and should be expected to be conducted in that manner.

13(d) subject to obtaining any necessary consent from a third party or parties, make such site visits and inspections as he considers appropriate, whether accompanied by parties or not.

The adjudicator has no authority to go on to other people's land without their consent. The site and other land used for the construction of the Works may never have been within or subsequently have passed out of the possession or control of one of the parties to the dispute.

The Scheme envisages that the adjudicator will visit the site either with both parties or neither. It is suggested that if one party objects the adjudicator should nevertheless go ahead with a site visit, informing all parties of the date and time he intends to be there, and then if any party fails to turn up, he should go ahead in the party's absence.

13(e) subject to obtaining any necessary consents from a third party or parties carry out any tests or experiments.

Such tests or experiments will be at the parties' expense. The adjudicator should therefore obtain the agreement of the parties to conduct such tests.

If any party fails to agree to a test or investigation considered necessary by the adjudicator the adjudicator may draw inferences from a refusal and find against that party on the basis that he has not proved his case.

13(*f*) *obtain and consider such representations and submissions as he requires, and, provided he has notified the parties of his intention, appoint experts, assessors or legal advisers.*

Obtain and consider such representations... is fairly straightforward. Having read the Referral Notice or any response provided by other parties the adjudicator may be unclear about some issues and require further evidence or argument. If the law is not clear, he may require submissions on that, in which case he may require simultaneous submissions to be provided by the parties within a fixed period.

The second half is less easy. *Having notified the parties of his intention appoint...* The adjudicator would be most unwise to make such appointments without the consent of at least one of the parties. He could be in difficulty if the parties objected on the grounds that such expert, assessor or legal adviser is not necessary. Such appointments will probably also require additional time and the parties may not agree to extend the 28 days.

If both parties are content for a matter to be decided without third party help, why should the adjudicator insist that it is necessary?

Where such advice is necessary, it would be better for the parties to agree upon the appointment of a third party and to agree to abide by his decision.

Adjudicators should not rush out to obtain legal advice merely because they are uncertain of a point of law. The parties and their advisers should do that.

Any advice that he gets should be given to the parties in sufficient time to enable them to comment on it. Where an adjudicator sought information from the Responding Party's subcontractors and the legal department of a local authority without informing the Referring Party the court refused to enforce the decision.[17]

17 *Woods Hardwicke Ltd* v. *Chiltern Air-Conditioning Ltd* [2000] BLR 23.

However, perhaps not too much reliance should be placed on this judgement because the adjudicator submitted a witness statement in support of the Referring Party in the enforcement proceedings!

13(g) give directions as to the timetable for the adjudication, any deadlines, or limits to the length of written documents or oral representations to be complied with, and ...

This is the hub of the Scheme. The adjudicator will almost always propose a deadline for the submission of any statement from the Responding Party. This, it is suggested, will normally be between 14 and 21 days to permit a week for reaching the decision and for any replies the Referring Party may wish to make.

Should the Referring Party consider that 14 days is too short to permit a proper defence to be provided, he should propose in the Referral Notice an extension of the period. Similarly, the Responding Party and the adjudicator, when they each receive the Referral Notice, should decide quickly whether either wishes to propose an extension and, if so, how long. The sooner timetabling problems are assessed and decisions made, the better for all parties.

... limits as to the length of written documents or oral representations to be complied with ...

In the first instance the length of the documents will be determined by the Referring Party in the Referral Notice. It is hardly possible to limit a response to less than the submission that it is responding to. There is equally little point in limiting a response to the length of the original submission — the time available will in any case limit what can be produced.

It is for submissions on law, or technical issues which arise during the course of the adjudication that limits will prove most useful.

13(h) issue other directions relating to the conduct of the adjudication.

This is an extremely wide power and would seem to give the adjudicator scope to order just about anything, provided that it complies with the other requirements of the Scheme and the law. Before using this power the adjudicator should consider:

(*a*) his duty to avoid incurring unnecessary cost;
(*b*) his duty to be impartial;
(*c*) the requirements of natural justice;
(*d*) the time restraints.

The power does, however, allow the adjudicator to act in an innovatory manner if he considers that that is in the interests of the parties.

The adjudicator is given times *within which* he is required to reach his decision. There is no reason why he should wait until the end of the period if he considers that he has all the information necessary to reach a decision.

Once the decision has been made the adjudicator has no power to change it substantially. It is therefore important that any replies to submissions of either party are made without delay, and any second thoughts are also conveyed immediately.

The courts have held that it is open to the adjudicator to correct his Decision provided that it is done forthwith.[18]

Summary

- The time periods start when the adjudicator receives the Referral Notice.
    - The Referral Notice should contain all the information the Referring Party wishes the adjudicator to have including both sides of the dispute.
    - The Referring Party should suggest in the Referral Notice any procedures he wishes to be adopted.

18 *Bloor Construction (UK) Limited* v. *Bowmer & Kirkland (London) Limited* and *Edmund Nuttall Limited* v. *Sevenoaks District Council.*

- The Responding Party should inform the adjudicator how long he requires to reply, and any other procedures he wishes to be adopted.
- The adjudicator:
  - will set the procedure for the adjudication when he receives the Referral Notice;
  - can take the initiative in ascertaining the facts and the law;
  - will make his decision in accordance with the contractual rights of the parties;
  - can take any failure to comply with his directions into account when making his decision;
  - may make his decision when he think he has enough information;
  - can, after informing the parties, seek to take advice from third parties.
  - Must make available to all parties any information that he obtains from people other than the parties.
- All correspondence with the adjudicator must be copied to all other parties.

7

General obligations of the parties

The Scheme imposes certain duties on the adjudicator and the parties which are in addition to the requirements of the procedure. It is important that the general duties are observed.

The adjudicator

Duties of the adjudicator

The adjudicator shall

(*a*) *act impartially in carrying out his duties and shall do so in accordance with any relevant terms of the contract and shall reach his decision in accordance with the applicable law in relation to the contract; and*

(*b*) *avoid incurring unnecessary expense.*[1]

The duty to act impartially imposes upon the adjudicator a duty to be fair to both parties in his dealings with them. It is also important the adjudicator is seen to be impartial and that means disclosing everything which could be perceived as possibly causing bias.

Personal friendship is not normally sufficient to indicate bias. It would normally be necessary to be able to imply some personal interest in the outcome. However, the adjudicator should draw the

1 Paragraph 12.

parties' attention to any dealings that he has ever had with the parties and they can take a view as to whether to complain or not.

It is also clear that the adjudicator must comply with any provisions in the contract which control adjudication. With adjudications under the Scheme, it is unlikely that the contract will have express provisions concerning adjudication because if it did they would probably be compliant and the Scheme would not be invoked. However, it is possible that the contract might extend the authority of the adjudicator by including disputes in connection with the contract which are outside the Act.

There would appear to be no reason why the parties cannot extend or modify the Scheme by agreement provided that they do not delete or change any of the eight compliance points.

'... *in accordance with the applicable law.*'

These words are significant because they partly define what adjudication is — it is a decision based upon the law. It is therefore a judicial process.

The adjudicator must ascertain the relevant law and if this is not forthcoming from the parties he must obtain it himself.

'... *avoid incurring unnecessary expense.*'

Although the adjudicator is given certain immunity from suit under paragraph 26, if he incurs unnecessary expense he will be in breach of his contract with the parties and could have his fees reduced by the amount of the unnecessary expense.

This general duty will provide argument for a party who does not like a course of action proposed by the adjudicator.

To decide all matters

The adjudicator shall decide the matters in dispute ...[2]

2 Paragraph 20. Opening statement.

The adjudicator is under an obligation to decide the matters. He is not entitled to say they are too difficult. He does no service to either the parties or the process of adjudication if he fails to give a decision.

Immunity to the adjudicator

The adjudicator shall not be liable for anything done or omitted in the discharge or purported discharge of his functions as adjudicator unless the act or omission is in bad faith, and any employee or agent of the adjudicator shall be similarly protected from liability.[3]

The adjudicator is protected from being sued by the parties unless he acts in bad faith. Bad faith is not a precise concept but would be very difficult to prove unless the offence was blatant such as fraud or failing to disclose a serious material matter, such as being the recipient of any sums of money that he decides should be paid by one party.

It would appear that this is a statutory immunity and will not protect the adjudicator from actions by third parties, which would require a wider immunity. This immunity is contractual since the scheme is an implied term of the contract.[4]

Such an immunity is necessary if the adjudicator is to start upon the adjudication without first agreeing terms with the parties. Without the immunity no-one would take on the job if there were serious contentious matters involved.

The parties

To comply with the directions of the adjudicator

The parties shall comply with any request or direction of the adjudicator in relation to the adjudication.[5]

This places an obligation upon the parties to comply. Failure to comply does not render the adjudication void and negate the

3 Paragraph 26.
4 s114(4) of the Act.
5 Paragraph 14.

adjudication in any way because the Scheme specifies what shall happen in the event of non-compliance.

The last five words ... *in relation to the adjudication* ... mean that the parties do not have to comply with any instruction that the adjudicator might (wrongly) give in relation to the work carried out under the contract.

One of the great worries, particularly amongst architects, has been that the adjudicator will influence what is actually constructed under the construction contract. That, of course, is not his function, which is to determine the contractual rights of the parties in relation to a dispute.

If the quality of work is in dispute it is not for the adjudicator to decide whether it is broken out and redone, but to decide whether or not the work as constructed complies with the specification. Whether or not it complies, it is always for the architect or engineer to decide with the client whether or not the work should be reconstructed.

It may be that, say, brickwork is condemned by an architect on the grounds of appearance. If the adjudicator decides that it complies with the standard required by the specification, it is for the employer, with the architect, to decide whether or not he is prepared to pay for the removal and re-erection of the work, not the adjudicator.

Failure to comply with an instruction

If, without showing sufficient cause, a party fails to comply with any request, direction or timetable of the adjudicator made in accordance with his powers, fails to produce any document or written statement requested by the adjudicator, or in any other way fails to comply with a requirement under these provisions relating to adjudication the adjudicator may:

- (*a*) *continue the adjudication in the absence of the party or of the documents or written statements requested,*
- (*b*) *draw such inferences from the failure to comply as circumstances may, in the adjudicator's opinion, be justified, and*

(c) *make a decision on the basis of the information before him attaching such weight as he thinks fit to any evidence submitted to him outside any period he may have requested or directed.*[6]

This is the general sanction, should a party not comply with any direction of the adjudicator.

The wording is very wide and includes any *request*. Thus if the adjudicator asks for further time either for himself or the other party and the requested party decides not to agree to such extension then the adjudicator can take that refusal into account under (*b*) when considering his decision. For instance, if the adjudicator asks for a test to be made and one side refuses, the adjudicator is entitled to ask himself why the party refused, and was it because he thought the results would be adverse to his case.

Confidentiality

The adjudicator and any party to the dispute shall not disclose to any other person any information or documents provided to him in connection with the adjudication which the party supplying it has indicated is to be treated as confidential, except to the extent that it is necessary for the purposes of, or in connection with, the adjudication.[7]

This is a general confidentiality clause. The intention is to protect parties from any unnecessary disclosure of documents which they are obliged to provide in the adjudication.

There is no general right of discovery as there is in an arbitration or court. The adjudicator has no authority to force disclosure of any document but he can ask for them and take any refusal into account when reaching his decision.

The parties cannot apply to the courts for the necessary orders to obtain disclosure. Nevertheless, parties should disclose any documents if requested to do so by the adjudicator unless they would be classed as 'privileged'.

---

6 Paragraph 15.

7 Paragraph 18.

Little is to be gained by withholding documents. Most documents will have to be disclosed if the matter goes to arbitration or to the courts through the process known as discovery. In addition, the adjudicator may *draw such adverse inferences from the failure to comply as the circumstances may, in the adjudicator's opinion, be justified.*[8]

8 Paragraph 15(b).

8

Settlement and resignation of the adjudicator

Procedure for a settlement

The primary aim of any dispute resolution process is to achieve a settlement. Parties who settle the dispute themselves are generally happier with the outcome than those who have a solution imposed upon them.

Adjudication is not the most conducive procedure for reaching a settlement but that will frequently be the outcome.

There are many reasons why a settlement may be an advantage to both parties. For instance:

(*a*) It improves the on-going relationships.
(*b*) It saves money.
(*c*) As the adjudicator's decision is final until changed by a court or arbitrator and the decision is enforceable in the courts, an adverse decision would be very expensive to reverse.

The most usual time for a settlement to be reached is after the Responding Party has submitted his response to the Referral Notice. The cases for the two parties are then clear and it is also apparent upon what basis the adjudicator will reach his decision. Unfortunately, the parties will have spent much of their costs involved in the adjudication by this time, but the adjudicator will not. His task really only begins after the submission of these statements.

It is to be hoped that a settlement will mean a saving of most of the adjudicator's fees and expenses.

Any settlement should be confirmed in writing so that the matter cannot be raised again in another adjudication. It should be signed by both parties.

When a settlement is reached, it will be necessary to stop the adjudicator immediately from doing any further work so that unnecessary cost can be avoided.

The Scheme indirectly addresses the situation:[1]

> *The parties to a dispute may at any time agree to revoke the appointment of the adjudicator. The adjudicator shall be entitled to the payment of such reasonable amount as he may determine by way of fees and expenses incurred by him. The parties shall be jointly and severally liable for any sum which remains outstanding following the making of any determination on how the payment shall be apportioned.*

Part of any settlement must therefore be an agreement to revoke the appointment of the adjudicator, and to decide how, by whom, and in what proportions the adjudicator's fees and expenses are to be paid.

The parties are jointly and severally liable for the adjudicator's fees which means that if his fees are not paid he may sue either party for the full amount of the fees and then that party would have to claim any over-payment from the other party. Even if one party has paid the full amount due from him, the adjudicator may still sue him for the other party's share. This could be unfortunate if one party goes into liquidation.

There is no standard form for recording the settlement but a number of points should be addressed in it.

(*a*) Identify the contract under which the dispute arose.
(*b*) Detail how the dispute arose.
(*c*) Give details of the dispute.

---

1 Paragraph 11(1).

(*d*) Give details of the settlement terms.
(*e*) Note the agreement that the adjudicator's appointment be revoked, and who will advise the adjudicator of it.
(*f*) Agree the apportionment of the adjudicator's fees and expenses.

Revocation of the adjudicator's appointment by the parties

There may be other reasons why the parties would like to revoke the adjudicator's appointment. These may include the following.

(*a*) The same dispute that occurred under two different contracts has been consolidated.
(*b*) There is a perceived possible clash of interest which the adjudicator indicates he does not see as a problem.
(*c*) The parties may become disillusioned with the adjudicator.
(*d*) Where the adjudicator has been appointed in the contract, the parties consider that his experience is inappropriate for a particular dispute.
(*e*) One of the parties wishes to engage the adjudicator as a consultant.
(*f*) The adjudicator has become ill.
(*g*) The time period is extended with the agreement of the parties so that it extends into a previous engagement of the adjudicator.

Doubtless there are many more.

The first case (*a*) is expressly covered by the Scheme.[2] For consolidation to occur, the party common to the two disputes must raise it at the time of the issue of the Notice of Adjudication, otherwise it will probably be too late.[3] The adjudicator will not know there are two similar disputes unless he is told. He cannot therefore propose consolidation.

When consolidation does take place one of the adjudicators will become redundant. It is up to the parties to decide which after they

2 Paragraphs 8(1) & (4).
3 See also Chapters 3 and 4.

have agreed to consolidate. The redundant adjudicator will be *entitled to be paid such reasonable amount as he may determine by way of fees and expenses.*[4]

In any of the other circumstances, the parties are free to revoke the appointment of the adjudicator but will be liable for his fees and expenses unless the revocation is attributable due to a default of the adjudicator.

> *Where the revocation of the appointment of the adjudicator is due to the default or misconduct of' the adjudicator, the parties shall not be liable to pay the adjudicator's fees and expenses.*[5]

It will be for the parties to prove that the adjudicator was in default.

Any agreement to revoke the adjudicator's appointment must include a statement as to how the adjudicator's fees are to be proportioned even if the parties believe that the revocation is due to the default of the adjudicator.

If the adjudicator should challenge any allegation of misconduct and the court agrees there will need to be an agreement as to how the fees should be proportioned, and if that is not done in the agreement to revoke the appointment there could be a further dispute.

The agreement should say: 'Should a court or arbitrator subsequently find that the adjudicator was not in default, any fees or expenses due to the adjudicator shall be proportioned and paid by the parties as follows ...'

Resignation of the adjudicator

The Scheme states that

> *An adjudicator may resign at any time on giving notice in writing to the parties to the dispute.*[6]

4 Paragraphs 8(4) & 25.
5 Paragraph 11(2).
6 Paragraph 9(1).

He must resign if a dispute is referred to him which is the same or substantially the same as one that has previously been referred to adjudication. This follows the normal legal practice that a matter cannot be tried twice.

> *An adjudicator must resign where the dispute is the same or substantially the same as one which has previously been referred to adjudication, and a decision has been taken in that adjudication.*[7]

Once a decision has been made the matter is said to be *res judicata*, i.e. a thing judged. Without this principle no dispute would ever be settled; one party would merely go on raising the matter in a different guise.

In any case, to allow a matter to be raised twice would offend the Act which says that the adjudicator's decision *is binding until the dispute is finally determined by legal proceedings, by arbitration (if the contract provides for arbitration or the parties otherwise agree to arbitration) or by agreement.*[8]

When the adjudicator resigns, a significant question for the parties is: 'are they liable for his fees?'

The Scheme states clearly two circumstances in which the adjudicator may resign and still be paid:

(*a*) when the dispute is the same or substantially the same as one already adjudicated;
(*b*) when he resigns because the dispute has varied significantly from the one referred to him in the Referral Notice.

It is silent about any other reasons the adjudicator may have for resigning and therefore implies that he will not be entitled to any payment:[9]

> *Where an adjudicator resigns in the circumstances referred to in paragraph (3), or where a dispute varies significantly from the*

7 Paragraph 9(2).
8 s108(3).
9 Paragraph 9(4).

dispute referred to him in the referral notice and for that reason he is not competent to decide it, the adjudicator shall be entitled to the payment of such reasonable amount as he may determine by way of fees and expenses reasonably incurred by him. The parties shall be jointly and severally liable for any sum which remains outstanding following the making of any determination on how the payment shall be apportioned.

The dispute may 'vary significantly' from that in the Referral Notice because of matters raised by way of defence by the Responding Party. For instance, it could be that a contractor believes that some of his work had been rejected on aesthetic grounds by the architect, and an architect is appointed as adjudicator. When the defence is submitted an additional ground for rejection is alleged, namely that the architect had been given advice by a structural engineer that the work was structurally unsound. It might then be that the appointed adjudicator, an architect, would consider himself unqualified to decide the issue.

The last point that it is necessary to consider is what happens after an adjudicator resigns.

The Scheme then requires that a new Notice of Adjudication shall be served and the adjudication starts again.[10]

(3) *Where an adjudicator ceases to act under paragraph 9(1):*
   (*a*) *the referring party may serve a fresh notice under paragraph 1 and shall request an adjudicator to act in accordance with paragraphs 2 to 7, and*
   (*b*) *if requested by the new adjudicator and insofar as it is reasonably practicable, the parties shall supply him with copies of all documents which they had made available to the previous adjudicator.*

The only difference, this time round, would be if the resigning adjudicator had been named in the contract. In those circumstances, the Referring Party would have to apply directly to the

10 Paragraph 9(3).

appointing body named in the contract, or if there was none, to an ANB.

The point about supplying the new adjudicator with the same documents that the previous adjudicator had is clearly sensible for saving both time and cost but it would probably be done even if the Scheme did not say so.

The alternative is for the adjudicator to take advice to help him over the problem, which of course he is able to do provided that he notifies the parties of his intention.[11] This would normally be the quicker and cheaper solution.

Summary

- An adjudicator may resign at any time.
- If he does he will be entitled to reasonable fees and expenses if
  - the dispute was one already adjudicated;
  - the dispute varied significantly so that he was no longer competent.
- The adjudicator must resign if the dispute has previously been adjudicated.
- The parties may revoke the adjudicator's appointment at any time.
- The adjudicator will then be entitled to reasonable fees and expenses, unless
  - the revocation was due to default or misconduct by the adjudicator.
- When the adjudicator resigns the Referring Party may serve a fresh Notice of Adjudication.

11 Paragraph 13(f).

9

The Decision

The Decision is the objective of adjudication, and all the actions of each party will be aimed at getting a decision in its favour.

The adjudicator's decision has to comply with the requirements of the Scheme for if it does not, an aggrieved party will seek to have it set aside.

It is therefore important that both the parties and the adjudicator are clear as to the context in which the decision is to be made.

The basis for the Decision

The adjudicator shall decide the matters in dispute.[1] The question is on what basis?

The adjudicator is required to act impartially *and shall reach his decision in accordance with the applicable law in relation to the contract.*[2]

Further he is given the authority *to take the initiative in ascertaining the facts and the law.*[3] The only purpose served by giving him this authority is if he is required to reach his decision on the basis of the facts and the law.

It is therefore clear that the adjudicator must decide the dispute in accordance with the legal rights of the parties arising from the contract.

---

1 Paragraph 20.
2 Paragraph 12(a).
3 Paragraph 13.

It must therefore be emphasised that it is not the adjudicator's job to give instructions, advice or directions on how the work is to be carried out or what should be built. It is his task to decide if the parties (or their representatives) have acted in accordance with the contract, or who, in accordance with the contract is liable when disputed events arise.

It will often be the case that the disputes detailed in the Notice of Adjudication are not in a form that the adjudicator can answer. He will then have to modify the questions so that they are in an acceptable form.

For instance, a dispute might arise if an architect condemns some brickwork on the grounds of its appearance, and instructs that it should be demolished. The contractor believes that it accords with the requirements of the specification. The dispute might then be referred to the adjudicator as, 'Should this wall be demolished or not?' That, of course, is not the right question, or even the question to which the parties require an answer.

The answer to that question would be 'yes' if the powers of the architect extended to ordering the removal of work. What the parties actually want to know is if the brickwork complies with the specification or not, because if it does the employer would have to pay for its demolition and re-erection and if it does not the contractor would.

Besides which, it is not within the power of the adjudicator to order the removal of work.

The adjudicator can decide only those disputes that are referred to him in the Notice of Adjudication. However, *he may take into account any other matters which the parties to the dispute agree should be within the scope of the adjudication or which are matters under the contract which he considers are necessarily connected with the dispute.*[4]

4 Paragraph 20 second sentence.

The first point to note is that the parties must agree for other matters to be considered which are not connected with the disputes notified in the Notice of Adjudication.

The most usual problem is when the Responding Party raises a counter-claim as a defence. For instance, a subcontractor refers a dispute on the valuation of the work he has done to the adjudicator. The main contractor, in his defence, alleges that the subcontractor has delayed him and he has been put to additional expense which he then has deducted from the sums due to the subcontractor.

Ignoring any arguments there might be about the right of set-off, the contractor has raised an issue which is not connected with the dispute referred to in the Notice of Adjudication. Whether the main contractor was delayed or not is another issue and may be another dispute. The adjudicator would have no authority to deal with that matter.

If additional unconnected issues are raised either in the Referral Notice or the Responding Party's response to it, then objection should be raised immediately because if the other party responds to it, it may be inferred that by so doing that party has agreed to the additional issues coming within the scope of the adjudication.

On the other hand the main contractor might allege that some of the work was deficient, in which case the issue of quality would be 'necessarily connected with the dispute' and the adjudicator could determine that issue.

Special authorities

While it is the job of the adjudicator to determine the contractual right of the parties, this is not helpful for all disputes which arise because certain actions taken by one of the parties to a contract may be right in the contractual context but wrong in substance. The main example is the issuing of certificates.

Where the contract states that one party's entitlement is dependent upon the certificate issued in accordance with a procedure laid down in the contract, that certificate will not, in the absence of express provision, be open to review by the courts.

It may be set aside, however, if it can be shown that the party issuing the certificate acted in breach of contract.[5] This might be very difficult when the matter challenged is one of opinion.

In order to be able to resolve this and certain other issues which arise in the construction industry, the adjudicator is given certain specific additional powers beyond just deciding the contractual entitlements.

The adjudicator *may open up, revise and review any decision taken or any certificate given by any person referred to in the contract unless the contract states that the decision or certificate is final and conclusive*.[6]

This is similar to the express power given to a court or arbitrator to look into the facts behind a certificate which is to be found in many of our standard arbitration or dispute resolution clauses. The adjudicator is given these powers under the Scheme and may therefore examine the facts upon which any such decision or certificates was based and substitute his own view for that of the person who made it.

This is particularly important in matters of payment or extensions of time where the client is himself making the decision or issuing the certificate. This is the situation which normally arises in construction subcontracts where the main contractor decides what the subcontractor should be paid and what extensions of time should be granted. It was the reason why Sir Michael Latham, in his report, proposed adjudication.[7]

The last words, *unless the contract states that the decision or certificate is final and conclusive*, would seem to negate the whole purpose of adjudication. All the paying party, who is generally the one who draws up the conditions for the contract, has to do is to state that all certificates, particularly interim certificates,[8] are final and conclusive and there will be no power for the adjudicator to

5 *Northern Regional Health Authority* v. *Crouch Company Ltd* [26 BLR 1] and *Balfour Beatty* v. *Docklands Light Railway* [78 BLR 42].

6 Paragraph 20(a).

7 Constructing the Team paragraph 9.5, p. 88 published by HMSO, July 1994.

8 See GC/Works/1 3rd Edition Conditions of Contract Condition 6(1), 18(3), 24(3), 26(3), 31(6), 39(2), 44(5), 50(3), 56(9).

upset such certificates unless it can be proved that the certificate was produced in breach of contract. That would normally be very difficult.

The origin of this exception would seem to be the JCT (Joint Contracts Tribunal) Contracts, which is the most commonly used form of contract for building works. Under these contracts the final certificate is final and conclusive evidence that the standard of materials and workmanship is to the reasonable satisfaction of the architect/contract administrator and the sum to be paid to the contractor.[9] However, under all these contracts the final certificate only becomes final and conclusive if arbitration or litigation proceedings have not been commenced with 28 days of issue of the certificate. This is a very different situation to that proposed by the Scheme. The difficulty for the draftsmen of the Scheme is that had they included a limited period for appeal in the Scheme they would have contravened the Act which gives a right to give notice of adjudication 'at any time'.[10]

The adjudicator can *decide that any of the parties to the dispute is liable to make a payment under the contract (whether in sterling or some other currency) and, subject to section 111(4)* of the Act, *when that payment is due and the final date for payment.*[11]

Whether this provision is, in fact, necessary may be open to debate because entitlement to payment is a legal right under the contract. It is useful for the purpose of clarity. The additional authority to determine when the sum is due and when the final date for payment is for the purposes of suspension of performance.[12]

One of the issues that could be referred to adjudication is whether an amount set-off has been correctly assessed. That situation is covered by s111(4):

> *Where an effective notice of intention to withhold payment is given, but on the matter being referred to adjudication it is decided that the*

9 JCT 80 Clause 30.9, JCT 81 Clause 30.8, IFC 84 Clause 4.7.
10 s108 (2)(a).
11 Paragraph 20(b).
12 See Chapter 12.

whole or part of the amount should be paid, the decision shall be construed as requiring payment not later than:

(a) seven days from the date of the decision, or
(b) the date which apart from the notice would have been the final date for payment, whichever is the later.[13]

The effect of this is to set the timetable for the exercise of the right to suspend performance.

The adjudicator may:

Having regard to any term of the contract relating to the payment of interest decide the circumstances in which, and the rates at which, and the periods for which simple or compound rates of interest shall be paid.[14]

The adjudicator is given power to award interest on any sums that he decides should be paid by one party to the other. When deciding the rate of interest the adjudicator is required to take into account any provisions in the contract relating to the payment of interest. Some standard contracts allow for the payment of interest on sums not paid[15] but others do not.[16] If the contract has no provision for interest the Late Payment of Commercial Debts (Interest) Act 1998 now provides for interest at 8% above base rate.[17]

If the Referring Party requires interest he should ask for it in his referral notice. The adjudicator will decide judicially what rate of interest should be paid and in doing this he will try to assess what the cost has been to the party receiving the payment of being out of his money for the period when it was not paid. However, if in the contract the parties have agreed a particular interest regime since this is a contractual entitlement the adjudicator will normally adhere to the contractual provisions when assessing interest due.

13 s111(4) of the Act.
14 Paragraph 20(c).
15 ICE Conditions of Contract, NEC, IChemE, GC/Works/1.
16 JCT Contracts.
17 Late Payment of Commercial Debts (Rate of Interest) (No. 2). Order 1998 SI 1998/2765.

Times for compliance with the Decision

The adjudicator may, in his decision, direct when the decision is to be implemented by the parties. It may be, for instance, that he will decide that a certain sum of money should be included in the next certificate issued by the employer. This would provide a simple means of implementation.

If he failed to make such a recommendation it would be interpreted that any additional payment would have to be made by a separate payment.

The Scheme says:

In the absence of any directions by the adjudicator relating to the time for performance of his decision, the parties shall be required to comply with any decision of the adjudicator immediately on delivery of the decision to the parties in accordance with this paragraph.[18]

There is, therefore, a default position should the adjudicator fail to specify a timetable.

Reasons

In the opinion of most commentators, reasons should not generally be given for the decision by the adjudicator. Reasons merely give the party who does not like the decision further grounds for having it set aside.

They also increase considerably the cost of the adjudication.

The absence of reasons does not mean that the adjudicator will not draft in rough the argument behind his decision. He must. It is not normally possible to check that all the points raised have been considered unless the arguments are reduced to writing.

However, the Scheme requires the adjudicator to give reasons if so requested by one party to the dispute:

If requested by one of the parties to the dispute, the adjudicator shall provide reasons for his decision.[19]

---

18 Paragraph 21.

19 Paragraph 22.

The extent of the reasons given will be a matter for the adjudicator but generally they will be minimal, and not comparable with those given by an arbitrator in his award.

The adjudicator, if asked, will give his reasons for his decision but not generally his reasoning. Reasons do not serve the same function in an adjudication as they do in arbitration, where they are required for an appeal. In adjudication it is not the decision that is appealed but the original dispute which is decided afresh. Reasons are therefore only given as an indication to the parties as to now the decision was arrived at.

The Scheme does not say when one of the parties should apply to the adjudicator if he wants reasons. Since the adjudicator has finished his task when he gives his decision and since there is no provision in the Scheme to amend the decision once it has been made, it would follow that any application must be made before the decision is reached.

In order to avoid unnecessary cost the parties should, if they really require reasons, ask for them with their respective statements. However, if one party objects to the request for reasons the cost of the reasons could be one of the factors that the adjudicator takes into account when he apportions his costs between the parties.[20]

Decision final

The Scheme contains no express term allowing the Adjudicator to correct errors in his Decision. However, errors have been raised in a number of cases.

If an adjudicator makes a mistake in his decision, whether in fact or law,[21] it will be enforced provided that he has not exceeded his jurisdiction,[22] that is because if he exceeds his jurisdiction he has not been authorised to make a mistake.[23]

20 Paragraphs 9(4), 11(1) & 25.
21 *SL Timber Systems Ltd* v. *Carillion Construction Ltd* [2000] 27 June.
22 *Bouygues (UK) Ltd* v. *Dahl-Jensen (UK) Ltd* (2000) 1 BLR 49 & CA.
23 *FW Cook Ltd* v. *Shimzu (UK) Ltd*.

In the absence of an express term it will be an implied term that the adjudicator can correct clerical or arithmetical errors in his decision provided that he does so within a reasonable time and without prejudice to the other party.[24,25]

Consequences of adjudicator exceeding the time limit for reaching the decision

The Scheme is very strict and requires the decision to be made within the prescribed time periods which are:[26]

The adjudicator shall reach his decision not later than

(*a*) *twenty-eight days after the date of the referral notice mentioned in paragraph 7(1), or*
(*b*) *forty-two days after the date of the referral notice if the referring party so consents, or*
(*c*) *such period exceeding twenty eight days after the referral notice as the parties to the dispute may, after the giving of that notice, agree.*

The adjudicator must comply with these periods because the Scheme appears to give him no latitude at all. Regrettably, it does not say whether a decision made a day late is a valid decision or not. It repeats similar words as are used when the adjudicator resigns:[27]

(2) *Where an adjudicator fails for any reason to reach his decision in accordance with paragraph 1:*
  (*a*) *any of the parties may serve a fresh notice under paragraph 1 and shall request an adjudicator to act in accordance with paragraphs 2 to 7; and*
  (*b*) *if requested by the new adjudicator and insofar as it is reasonably practicable, the parties shall supply him with copies of all documents which they had made available to the previous adjudicator.*

---

24 *Bloor Construction (UK) Ltd* v. *Bowmer & Kirland (London) Ltd* [2000] BLR 764.
25 *Edmund Nuttall Ltd* v. *Sevenoaks District Council* [unreported].
26 Paragraph 19(1).
27 Paragraph 19(2).

These words indicate that the Referring Party has to start again (because the paragraph 1 referred to is paragraph 1 of the Scheme) and issue a new Notice of Adjudication. This is a very onerous procedure and likely to cause considerable delay.

Most other procedures state that a late decision is valid unless a new adjudicator has been appointed.[28]

It is suggested that even if the decision is overdue by a few days, the parties should agree to accept it and save themselves from unnecessary further expense.

The form of the Decision

The Scheme gives no guidance on what form the decision should take. There are certain matters that must be included in the decision.

The written decision may have two different uses:

(*a*) For implementation by the parties, and
(*b*) for enforcement by the courts should one party fail to implement it.

The decision must therefore cover both these aspects.

Implementation by the parties

The parties will want to know precisely what is covered by the decision. If the decision includes the valuation of any work, it is essential that it is clear exactly what is covered in the evaluation so that double accounting does not occur.

This means that it has to be explained in terms of the payment mechanism of the contract. For instance, when there is a lump sum for an element of work, and a part of that element is varied the adjudicator must make clear whether his valuation includes or excludes any credit that may be involved for work omitted by the variation. Otherwise it will be impossible to determine the final price.

28 See ICE Adjudication Procedure, and ORSA Adjudication Procedure.

Enforcement by the Court

This is a matter which should not arise but, like all contracts, the adjudication needs to be written to cover all the matters necessary should legal proceedings follow.

This will require certain additional information beyond that necessary for the sole use of the parties. This additional information will include details of:

(*a*) the contract under which the adjudication takes place;
(*b*) the circumstances which led to the particular reference to the adjudicator;
(*c*) the procedures adopted by the adjudicator;
(*d*) the information upon which the adjudication is based.

The Contract. There is little need to do more than to refer to the Contract, giving details of the parties, the date it was formed and its purpose, the manner and timing of the appointment as adjudicator, and the date of the Notice of Adjudication and the date of service of the Referral Notice on the adjudicator.

The circumstances of the dispute. The background to the dispute should become apparent from the submissions of the parties, and a resumé of the facts concerning the dispute should be given in sufficient detail to enable the reader to understand the differences between the parties. It will also demonstrate that the adjudicator has grasped the essential points concerning the dispute.

It is not necessary for the adjudicator to set out in full detail all the facts in a manner similar to the reasoned award of an arbitrator or the judgement of a court. If legal action follows, all the surrounding facts will then be given in evidence. It is not the adjudicator's decision that is referred to the court or arbitration but the original dispute.

What the court requires to know is sufficient about the dispute to be able to implement the decision.

However, to date around 25 Decisions have not been enforced by the Courts. This is around 0.25% of the adjudications that have taken place. The chances of resisting enforcement are therefore very low.

The Procedures adopted by the adjudicator. This is one of the most important sections. Since the adjudicator's appointment is contractual and he is required to act in accordance with the Scheme, it is important for him to demonstrate that the procedures he adopted comply with the requirements of the Scheme, and, where appropriate, the contract.[29] Most applications for setting-aside the decision are likely to be on the grounds that the adjudicator has not complied with these requirements, or has improperly exceeded them.

The adjudicator should state precisely:

(*i*) the submissions that he has received from the parties, or others;
(*ii*) any further requests for information that he has made;
(*iii*) any meetings that he has held with the parties or others together with any requests and agreements concerning these meetings;
(*iv*) any advice that he has sought, either with or without the parties' specific agreement;
(*v*) the nature of that advice and the extent to which the adjudicator has accepted it;
(*vi*) any visits to Site he has made;
(*vii*) any other actions that he has taken.

Generally, the adjudicator will state when the procedures or actions taken were with the agreement of the parties, and when they were at his own initiative.

29 Paragraph 12(a).

The Information upon which it is based. It is essential from the adjudicator's point of view that all information used in the adjudication is listed in case the decision is challenged and particularly on the grounds that the adjudicator overlooked some important piece of information.

The adjudication can be based only upon the information supplied and the adjudicator is under no duty to make investigations. He may ask for information to clarify the submissions.[30] If the parties fail to provide the adjudicator with vital information that is a matter for them.

The information given under (*iii*) above will cover most of the information but there may be documents submitted by the parties which are not listed in their submissions, and there may be documents which were in the adjudicator's possession prior to the particular reference to which he may have referred. A comprehensive list of all documents referred to should be attached to the adjudication, possibly as an Appendix.

If reasons for the decision are requested by one[31] or all the parties the adjudicator should include them after detailing the information on which his decision is based.

The reasons will generally be restricted to his decisions in respect of the facts and the law and will not give detailed explanation as to how he arrived at those findings.

If adjudication is to perform its purpose the decision must be enforced, and if a party objects, the correct course will be to refer the matter to arbitration or litigation, not to have it set aside pending final resolution. The aim of the adjudicator will be to minimise the opportunities for having the decision set aside.

For that reason the adjudicator will minimise the extent of the reasoning given for his decision.

30 Paragraph 13.
31 Paragraph 22.

The Decision. The court will want to know precisely what has got to be enforced as a result of the decision. Generally, matters going to the court will concern failure to pay an amount directed by the adjudicator.

The court will require identical information to the parties, except the court will not have the background knowledge concerning what has already been paid and how those sums had been arrived at. Generally, that will not be a problem since the amount assessed by the adjudicator will be a sum independent of other amounts certified under the contract.

Summary

- The adjudicator must decide all the matters in dispute.
- The adjudicator must reach his decision within the allotted time.
    - If he fails to do so the Referring Party may issue another Notice of Adjudication;
    - the adjudicator will not be entitled to any payment.
- The adjudicator can only decide matters not in the Notice of Adjudication if
    - the parties agree;
    - they are necessarily connected to the dispute.
- The adjudicator may open up, revise and review any decision or certificate issued under the contract unless
    - the decision is stated to be final and conclusive.
- If requested by one of the parties, the adjudicator must give reasons for his decision.
- An adjudicator can correct errors in his decision provided it is done quickly.

10

The effects of the Decision

The decision of the adjudicator shall be *binding upon the parties and they shall comply with it until the dispute is finally determined by legal proceedings, by arbitration (if the Contract provides for arbitration or the parties otherwise agree to arbitration) or by agreement between the parties.*[1]

The adjudicator's decision is therefore a temporary decision which is put into effect. It can, however, be replaced by a subsequent decision of a court or arbitrator. The parties may also agree to vary it.

Nothing within the Act or the Scheme removes the rights that are available under the contract to vary the works constructed under the construction contract. Should an adjudicator give a direction (erroneously) that directs either how the works should be carried out or whether work should be broken out or remain, there is nothing which prevents a person empowered to give instructions under the contract from issuing an order (provided that it is within powers bestowed by the contract) which overturns what has been directed by the adjudicator. The order will then be valued in accordance with the requirements of the contract taking into account the decision of the adjudicator.

It should be noted that it is the dispute which is referred to the legal process, not the adjudicator's decision. This is an important distinction because it means that the courts or arbitrator will review

---

1 Paragraph 23(2) reflecting s108(3) of the Act.

all the evidence afresh and not rely solely upon that which was put to the adjudicator.

The parties are therefore free to present their cases in any manner they wish. It does not matter if they omit some details. It may reduce the likelihood of success but it will not prevent the use of such details in a subsequent appeal.

It also means that it is not necessary for the adjudicator to give reasons for his decision, although the Scheme requires him to do so if requested by one of the parties,[2] as these are not needed to assist the court at an appeal as they would be in the case of an arbitrator's award.

Compliance with an adjudicator's decision is essential if adjudication is to achieve its purpose.

Under standard contracts this will generally not be a problem. Most decisions will involve the payment of money and/or grants of extensions of time. Under standard contracts decisions on these matters will become contractual obligations which the certifier (architect, engineer or quantity surveyor) will be obliged to certify. A certifier would clearly be in breach of duty if he failed to do so. This could put the employer into breach of contract.

Under non-standard contracts, which is the more likely for adjudications under the Scheme, the position may be more difficult because there may be no independent certifier, and the paying party will decide the amount due. In effect, he decides what he is prepared to pay.

Unfortunately, the draughtsmen of the Act did not consider properly the manner in which an adjudicator's decision was to be implemented. They provided that the Arbitration Act 1996 might be used, but did not consider how.

The chosen method is clumsy and not readily understandable.

The Act permits the Scheme to use the Arbitration Act 1996 as the means of enforcement.

2 Paragraph 22.

For England and Wales the Scheme may apply the provisions of the Arbitration Act 1996 with such adaptations and modifications as appear to the Minister making the scheme to be appropriate.[3]

The Scheme adopts only one Section, s42, of that Act for the purposes of enforcing the adjudicator's decision. This section deals with the enforcement of procedural orders and directions of the Arbitrator, not his award. It is intended for use during the interlocutory stages, that is while the parties are defining their respective cases. It is not for the enforcement of an Award.

The Scheme does not directly tie the decision into this section. It states:

In his decision. the adjudicator may, if he thinks fit, order any of the parties to comply peremptorily with his decision or any part of it.[4]

A non-lawyer would be forgiven for not realising immediately what that means or implies. In order to understand it fully it is necessary to read s42 of the Arbitration Act 1996 which has been fortunately well drafted and is readily understandable.

The Scheme is permitted to make adaptations and modifications[5] to the Act and these have been substituted in the version of s42 which is given below. The adaptations are

Section 42 – Enforcement of Peremptory Orders of Tribunal

42(1) Unless otherwise agreed by the parties, the court may make an order requiring a party to comply with a peremptory order made by the tribunal.

(2) An application for an order under this section may be made:

(a) by the ***adjudicator*** *(upon notice to the parties),*

(b) by a party to the ***adjudication*** *proceedings with the permission of the* ***adjudicator*** *(and upon notice to the other parties), or*

(c) This section has been deleted.

3 s108(6).

4 Paragraph 23(1).

5 Paragraph 24.

(3) **This section has been deleted.**

(4) No order shall be made under this section unless the court is satisfied that the person to whom the tribunal's order was directed has failed to comply with it within the time prescribed in the order or, if no time was prescribed, within a reasonable time.

(5) The leave of the court is required for any appeal from a decision of the court under this section.

It is, of course, the first line of sub-section (1) upon which the Scheme relies. *The court may make an order requiring a party to comply with a peremptory order made by the tribunal.*

The Scheme seems to leave it up to the adjudicator to decide whether or not to order a party to comply peremptorily with his decision. This is unfortunate because should he decide not to, the only way to get a decision enforced should a party fail to comply would be to start another adjudication. The second adjudicator would then probably be persuaded to order his decision to be complied with peremptorily.

It would seem, therefore, that adjudicators would normally order the parties to comply with the decision peremptorily. Routinely making such an order will avoid any innuendo that he considers that a particular party is unlikely to comply with his order.

It would seem wise for parties to ask for an order in that form.

The exact effect of sub-section (3) is less clear. Does failure to comply with an adjudicator's decision itself become a dispute and therefore referrable to adjudication? If so, it would seem that this enforcement procedure will be lengthy. If failure to comply is not adjudicable then it is hard to see what possible effect this sub-section can have because the last act of the process is for the adjudicator to give his decision.

Finally, it should be noted that leave of the court, that is permission of the court, will be required to appeal against any order of the court in relation to enforcement of the adjudicator's decision. Again, it is hard to see what possible grounds of appeal would succeed.

The only powers of the court in relation to an adjudicator's decision are either to enforce it or to set it aside. They will enforce it if they think that it has been arrived at properly, and will set it aside if it has not. Since it is only a temporary decision, the parties' actual safeguard is that the dispute itself is referred to the court or an arbitrator.

Generally, it would seem that the courts will take the view that if there is to be an appeal it should be on the dispute rather than the adjudicator's decision.

Whether this procedure will work or not is an interesting legal problem but not one for this book.

The procedure to be adopted under the Scheme to implement a decision if a party is unwilling to comply will be:

(*a*) The aggrieved party must check that the adjudicator has ordered the other party to comply peremptorily with his decision.
(*b*) If he has, application should be made to the court for enforcement. This will normally involve the employment of a solicitor, who will make the necessary application.
(*c*) If the adjudicator has failed to order a party to comply peremptorily then the aggrieved party will have two choices:
    (*i*) to apply to the courts and seek summary judgement, or
    (*ii*) to start a new adjudication asking the second adjudicator to order the other party to comply peremptorily with his order.

There is one decision the adjudicator might make which has a different effect. That concerns a party's right to suspend performance if payment is not made by the *final date for payment.*[6] This is explained fully in Chapter 12.

Essentially, if a payer is not going to pay the full amount due because, for instance, he has some counter-claim which he wishes to set-off against the amount due, he is required to issue a notice to

6 s112 & s113 of the Act.

that effect a prescribed period before the final date for payment. This notice is defined as an effective notice.

If the payee wishes to challenge the right to set off such sums, he will refer the matter to an adjudicator. If the adjudicator decides that the payer should not have made the deduction, the sum wrongly deducted becomes due seven days after the decision.[7] The payee can then give seven days' notice that if he is not paid the amount within the following seven days he will suspend performance of the contract. This requires no court intervention — he merely stops work.

Time will show how willing the courts will be to enforce such peremptory orders. The courts are supportive of arbitrators. There is no reason to believe that they will not be equally supportive of adjudicators.

If the courts are supportive the adjudication system will serve its purpose, but if they prove reluctant to act the adjudication system will need a rethink.

Summary

- The parties must comply with the decision of the Adjudicator.
- The decision is binding until the dispute is finally resolved by arbitration, litigation or agreement.
- The adjudicator may order the parties to comply peremptorily with his decision:
  - failure to comply with a peremptory order permits the courts to enforce the decision.
- The adjudicator can state a final date for payment in his decision:
  - failure to pay by the final date for payment entitles a party to issue a notice of intention to suspend performance;
  - the payer then has seven days in which to pay in full.

7 s111(4)(a).

11

Payment of the adjudicator

It is not essential for the parties to agree a fee with the adjudicator. The Scheme provides that the adjudicator is *entitled to the payment of such reasonable amount as he may determine by way of fees and expenses reasonably incurred by him.*[1] It is the adjudicator who initially determines the amount of the fees and expenses.

If one or all parties consider that the adjudicator's fees are unreasonable the solution is not to pay them and force the adjudicator to sue for them. In that case the courts will probably require the full amount of the fees to be paid into court before they will consider the action. The loser will normally be required to pay the costs of the other side. Such a course should be embarked on only after considerable thought, otherwise the costs incurred may exceed the over-charge element of the adjudicator's fees.

There will be only limited scope for challenging the amount charged by the adjudicator since it will be up to the party who challenges the fees and expenses to prove to a court that the amount charged was unreasonable.

However, there is nothing in the Scheme which prevents the parties and the adjudicator from agreeing the rates to be charged by the adjudicator or the method of assessing them.

If the adjudicator is named in the contract the parties can negotiate a basis for charging before or immediately after the

1 Paragraph 25.

contract has been formed. The fees can therefore be negotiated either as a lump sum or on a time-spent basis.

When the adjudicator is neither named in the contract nor appointed before a dispute arises, it is impractical to agree the adjudicator's terms prior to appointment because of the time limits set out in the Scheme.

Clearly, the reason why the Scheme does not require agreement of fees before the adjudication starts is so that no party or the adjudicator can delay the adjudication by insisting on such agreement.

The adjudicator may, in his decision, direct how his fees and expenses are to be proportioned between the parties.

The Scheme states: *The parties shall be jointly and severally liable for any sum which remains outstanding following the making of any determination on how the payment shall be apportioned.*[2]

This rather curious wording makes the parties jointly and severally responsible for the adjudicator's fees and expenses. It means that if one party fails to pay the proportion of the fees allocated to him by the adjudicator, the adjudicator can sue any other party to the dispute for the amount not paid. That party will then have, in turn, to sue the defaulting party to recover the amount that he has overpaid.

This is the only safeguard that the adjudicator has for the payment of his fees under the Scheme. It does not give him a lien on the decision, and does not permit him to withhold it until he is paid.

As soon as possible after he has reached a decision, the adjudicator shall deliver a copy of that decision to each of the parties to the contract.[3]

It would therefore be a default by the adjudicator if he attempted to withhold his decision until he received payment.

2 Paragraph 25.
3 Paragraph 19(3).

Payments if the adjudicator resigns

The *adjudicator may resign at any time on giving notice in writing to the parties to the dispute.*[4] No time period is stated for this notice which will therefore be effective when it is received by the parties.

The adjudicator will be entitled to payment only if he resigns for one of two reasons:[5]

(*a*) The dispute is one that has already been adjudicated, or
(*b*) the dispute differs substantially from the one referred to him in the Referral Notice and because of that he decides that he is not competent to decide it.

In these circumstances *the adjudicator shall be entitled to the payment of such reasonable amount as he may determine by way of fees and expenses reasonably incurred by him. The parties shall be jointly and severally liable for any sum which remains outstanding following the making of any determination on how the payment shall he apportioned.*[6]

It should be noted that the adjudicator will be the claimant for his fees in this situation and will have the burden of proving he is entitled to them.

He must show both that the dispute differs significantly from that defined in the Referral Notice *and* that those differences render him incompetent to decide it.

It is curious why the draughtsmen chose the Referral Notice as the basis for the adjudicator to decide whether the dispute has varied so much that he has become incompetent to decide it. It was not the Referral Notice upon which the adjudicator based his decision to accept the appointment but the Notice of Adjudication.

However, the Referral Notice must only cover the matters raised in the Notice of Adjudication and the adjudicator may refuse to consider them because it is the Notice of Adjudication that defines his jurisdiction, and it is upon that Notice he based his decision to

4 Paragraph 9(1).
5 Paragraph 9(4). These items are discussed in Chapter 8.
6 Paragraph 9.

accept the appointment. The Scheme says he may consider additional disputes with the consent of the parties. It does not require him to do so.[7]

The thinking must be that it will be the arguments of the receiving party when he submits his defence that may raise matters outside the competency of the adjudicator.

In other situations when the adjudicator resigns it would appear that he is not entitled to any payment.

When the parties revoke the adjudicator's appointment

The parties to a dispute may at any time agree to revoke the appointment of the adjudicator.[8]

Such revocation requires the agreement of the parties to the dispute, and that may not always include all the parties to the contract or it may involve more if the dispute has been joined with a similar one under a different contract. A referring party cannot unilaterally terminate the adjudication.

The most common reason for revoking the appointment will be that a settlement has been reached. The aim of adjudication is to assist the resolution of disputes and the process will have achieved its aim if a settlement is reached.

This should be a common occurrence. Experience has shown that many arbitrators have been appointed in the past merely to break an impasse. Only about 10% of references to arbitration actually proceed to a hearing, because a settlement has been reached earlier. The same is true of actions in the courts. There is, therefore, no reason to expect that adjudication will be any different.

Naturally, if the parties revoke his appointment the adjudicator will be entitled to payment for any work done. The Scheme so provides: ... *the adjudicator shall be entitled to the payment of such*

7 Paragraph 8.
8 Paragraph 11(1).

reasonable amount as he may determine by way of fees and expenses reasonably incurred by him. The parties shall be jointly and severally liable for any sum which remains outstanding following the making of any determination on how the payment shall be apportioned.[9]

When a settlement is reached the parties should, as part of their agreement to revoke the appointment, agree how the fees should be apportioned. Normally, this will be what happens and in that case it would be expected that they would so tell the adjudicator when they inform him of the revocation of his appointment. The adjudicator would then be expected to abide by that decision.

The wording of the Scheme in respect of the adjudicator's entitlement to payment when his appointment is revoked is similar to that used in the case of the adjudicator resigning for reasons which still entitle him to payment.[10] He determines his own reasonable fees and expenses and determines the apportionment of them between the parties.

Reasonable fees and expenses

Parties are always going to wish to know the likely cost of an adjudication before embarking on it.

The adjudicator is not normally likely to be involved in much work before the statements of all the parties to the dispute have been submitted to him.

He will have to read the Referral Notice carefully and decide the timetable and other arrangements for the adjudication. The time taken will vary according to the length of the Referral Notice. He may also decide that he wishes further information from the referring party pending the submission of the other party's statement.

It is unlikely that he will be involved in more that one or two days' work before all the statements are with him. That will

9 Paragraph 11(1).
10 Paragraph 9(4).

normally be between two and three weeks after receipt of the Referral Notice.

Once all the statements are in, the adjudicator will then prepare his decision. The time taken will increase if one of the parties asks for reasons.[11] Reasons may be expected to add two or three days to the cost of preparing the decision.

Clearly, the time taken will depend upon the complexity of the dispute but unless an extension of time is agreed the adjudicator cannot take more than about a week to prepare his decision. This means that normally it can be expected that the adjudicator will spend between three and eight days working on the adjudication.

Normally, adjudicators will be senior members of their profession and will expect to be paid at a rate commensurate with that position.

For a rough guide to what might be a reasonable fee the parties can inquire the going rate within the appropriate profession for senior people. This will not be conclusive because it will be for the court to decide what is reasonable. and the scale of fees with which judges are familiar will be the going rates in the legal profession. A quick check on the local rates for solicitors will provide another guide.

The rates will vary considerably between adjudicators. However, the lowest hourly rate may not work out the cheapest overall. People who are able to charge high rates can do so because they manage to work very quickly and not all adjudicators work at the same speed.

Actual rates may vary from £75 to over £300 per hour. The Scheme does not lay down any standard, nor will most ANBs, for to do so may well contravene the Restrictive Practices Acts.

The level of expenses will again be determined by the complexity of the issues in the dispute. The adjudicator's major expenses will be:

(*a*) travel if a site visit or other inspection is required;
(*b*) travel if a meeting is required;
(*c*) advice, either technical or legal.

11 Paragraph 22.

All these matters are largely controllable by the parties. Generally, it will be the parties who suggest a site visit or meeting. However, if the adjudicator does, then if either party considers it unnecessary he should say so immediately it is proposed.

The adjudicator is required to inform the parties before taking third party advice. If either party considers it is unnecessary he should contact the other party. If they both (or all if there are more than two) agree that it is unnecessary they should inform the adjudicator of their agreement. The adjudicator should then not obtain the advice but decide on the information available to him.

Thus the parties should be able to make a fair assessment of what the adjudication will cost because they do not have to consider what the expenses of the other party are likely to be.

The Adjudicator has no power to apportion the costs of the parties themselves. Each must pay his own costs. Again the control of the cost is with the parties.

If one party decides to spend a large amount on his case that will only indirectly affect the other party.

Unless the parties and the adjudicator have agreed the basis upon which the fees and expenses are to be calculated, any party may subsequently wish to challenge the amount of the fees in the courts, however, the courts may consider the rate charged but will not substitute its own ideal of the number of hours spent for that of the adjudicator in the absence of bad faith.[12]

Before doing so it is advisable to determine what is a reasonable fee for the adjudicator. In doing so some margin should be allowed over and above the rate that is perceived to be reasonable because judges will probably give the benefit of the doubt to the adjudicator.

Summary

- The adjudicator will normally be entitled to be paid his fees and expenses reasonably incurred.

12 *Stubbs Rich Architects* v. *WH Tolley & Sons Ltd* (8 August 2001).

- The adjudicator can allocate the proportion of his fees and expenses that each party bears.
- If the adjudicator resigns he will be entitled to payment if he resigns because:
  - the dispute is similar to one that has already been referred to adjudication, or
  - the dispute varies significantly from the one in the referral notice and consequently he feels incompetent to deal with it.
- If the adjudicator resigns for other reasons he is not entitled to any payment.
- The parties may revoke the appointment of the adjudicator at any time:
  - If his appointment is revoked he is entitled to fees and expenses reasonably incurred.
  - The adjudicator may apportion such fees and expenses between the parties.

12

The payment provisions

The payment provisions of the Act have probably had a greater impact upon the construction industry than the adjudication provisions. They have forced paying parties to address the costs of variations and delay claims as applications for payment are made and not to delay consideration of them by prevarication. They have had the effect that Sir Michael Latham intended.

The Housing Grants, Construction and Regeneration Act 1996 provides certain rights to parties to a construction contract in respect of payment. These are:

(*a*) for payment by instalments (or stage payments) for contracts over 45 days' duration;
(*b*) to be informed of the amount to be paid in any instalment and when it is due for payment;
(c) to be given notice if it is intended that any payment be withheld;
(*d*) the right to suspend performance if payment is not made within a specified period;
(*e*) the outlawing of 'pay when paid' clauses.

The only sanction in the Act for failure to pay on time is the right to suspend performance, i.e. stop work on the contract. There is no general right under English Law to interest on overdue payments unless the contract provides for such payments.

If a contract is silent on interest it will be very difficult to obtain any interest on sums paid late under a contract with a complying adjudication provision. If the matter is referred to the adjudicator it is unlikely that the adjudicator will award interest because of this wording. It is the adjudicator's task to establish the contractual rights of the parties and if they have no contractual right to interest it is suggested that the adjudicator has no right to make a decision including it.

However under the Scheme, an adjudicator is entitled *having regard to any term of the contract relating to the payment of interest decide the circumstances in which, and the rates at which, and the periods for which simple or compound rates of interest shall be paid.*[1]

It is suggested that this gives the adjudicator the power to award interest even when the contract is silent. These words would be unnecessary it this was not so. It would seem that their intention is to require the adjudicator to apply the interest rate in the contract if there is one specified, but if there is not then he can award such interest as he considers appropriate.

The rate in most construction contracts where one is specified is between one and three percent above base rate. However, recently standard contracts have been increasing this figure and figures between 5% and 10% above base rate are not uncommon. Interest in accordance with the Late Payment of Commercial Debts (Interest) Act 1998 is regularly requested although the Act only applies to companies who have less than 50 employees. The argument is that this is a reasonable rate recognised by the government.

The requirements of the whole of this part of the Act are aimed solely at the right to suspend performance.

These requirements are covered in the Scheme in **Part II – Payment.**

Because the main sanction is suspension of performance, which the courts would consider a drastic step, the Scheme has been drafted with enforcement in mind. It deals with all these matters but is very badly drafted from the user's viewpoint.

---

1 Scheme Part I paragraph 20(c).

Most standard contracts have been amended to provide for these items and the Scheme is most likely to apply to those contracts which are formed by an exchange of letters, or entered into orally.

In those circumstances reliance upon the Scheme should not lead to much trouble provided that two principles are adhered to:

(*a*) that valuations are made at the end of every month and payment is made in full within a month thereafter;
(*b*) that if any sums of money are to be withheld for any reason a notice stating the amount to be withheld is given within three weeks of the end of the month for which payment is due.

Entitlement and amount of stage payments

The Act says that *a party to a construction contract is entitled to payment by instalments, stage payments, or other periodic payments for any work under the contract unless:*

(*a*) *it is specified in the contract that the duration of the work is to be less than 45 days, or*
(*b*) *it is agreed between the parties that the duration of the work is estimated to be less than 45 days.*[2]

It has been suggested that some employers or contractors will specify work in small amounts in order to avoid these provisions, but any such evasion is unnecessary because *the parties are free to agree the amounts of the payments and the intervals at which, or the circumstances in which, they become due.*[3]

The Act therefore gives considerable freedom to the parties to negotiate the payment mechanism. It allows for a wide variety of payment regimes including traditional monthly instalments, milestone payments, stage payments, payments according to a graph, and others.

2 s109(1).
3 s109(2).

This freedom does give rise to means of total avoidance. For instance, the interval period could be made as long as or longer than the contract period.

In the absence of agreement between the parties *the relevant provisions of the Scheme for Construction Contracts apply.*[4]

When a contract contains no provisions for stage payments the Scheme provides for periodic payments at specified intervals. Each interval is defined as a *relevant period.*[5]

For the purposes of the Scheme the *'relevant period' is defined as a period which is specified in, or is calculated by reference to the construction contract or where no such period is so specified or is so calculable, a period of 28 days.*[6]

This would indicate that mechanism for payment in the contract is divisible into its components. If there is an adequate mechanism for defining the time when payment is to be made then that contract mechanism is adhered to when the relevant period is being defined.

So if milestones are defined but there is an adequate mechanism for valuing the work at each milestone, those milestone dates will suffice for the purposes of defining the relevant period but reference will have to be made to the Scheme for the means of valuing the payment due.

Mechanism for determining the sum due

The Act requires that *every construction contract shall provide an adequate mechanism for determining what payments are due under the contract, and when.*[7]

There is no indication in the Act as to what constitutes an adequate mechanism. It will be for the courts or an arbitrator to decide. The Scheme can therefore possibly be used as a yardstick by

4 s109(3).
5 Scheme Part II paragraph 12.
6 Scheme Part II paragraph 12.
7 s110(1).

draughtsmen of contracts when they draw up these provisions. Clearly, the mechanism in all our current standard contracts is adequate — it is generally certification by a third party.

The Scheme for the purposes of these payment provisions calls contracts which either specify or the parties agree that the duration will be over 45 days as *relevant contracts.*[8] *Relevant contracts* are therefore ones for which instalment payments are required.

Paragraph 1 of Part II of the Scheme states:[9]

Where the parties to a relevant construction contract fail to agree:

(*a*) *the amount of any instalment or stage or periodic payment for any work under the contract, or*
(*b*) *the interval at which, or the circumstances in which, such payments become due under the contract, or*
(*c*) *both of the matters mentioned in sub-paragraph (a) and (b) above the relevant provisions of paragraphs 2 to 4 below apply.*

The Scheme is of course more detailed than that simple explanation. The valuation at the end of the relevant period is defined as *the aggregate of the following amounts:*[10]

(*a*) *an amount equal to the value of any work performed in accordance with the relevant construction contract during the period from commencement of the contract to the end of the relevant period (excluding any amount calculated in accordance with paragraph 2(b)),*
(*b*) *where the contract provides for payment or materials, an amount equal to the value of any materials manufactured on site or brought onto site for the purposes of the works during the period from the commencement of the contract to the end of the relevant period, and*
(*c*) *any other amount or sum which the contract specified shall be payable during or in respect of the period from the commencement of the contract to the end of the relevant period.*

8 Scheme Part II paragraph 12.
9 Scheme Part II paragraph 1.
10 Scheme Part II paragraph 2(2)

The value of work means *an amount determined in accordance with the construction contract under which the work is performed or where the contract contains no such provision, the cost of any work performed in accordance with that contract together with an amount equal to any overhead or profit included in the contract price.*[11]

This allows for the situation where there are no rates or prices contained in the contract and will cover varied work. It begs the issue as to how it is to be valued if there is no price in the contract from which to derive overheads and profit. These will have to be agreed at the time or determined by the adjudicator.

Work means *any work or services mentioned in section 104 of the Act.*[12]

This is very clear and straightforward. It does not include any materials which are still in the manufacturer's or subcontractor's works even if the contract provides for payment of such materials or things.

The amount of any payment by way of instalments or stage or periodic payments in respect of a relevant period shall be the difference between the amount[13] *calculated above and the aggregate of any sums which have been paid or are due for payment by way of instalments, stage or periodic payments during the period from the commencement of the contract to the end of the relevant period.*[14]

There is then a cap on the total amount to be paid so that, for instance, the over-ordering of materials will not result in over-payment.

The amount to be paid *shall not exceed the difference between*[15]

(*a*) *the contract price, and*
(*b*) *the aggregate of the instalments or stage or periodic payments which have become due.*

11 Scheme Part II paragraph 12.
12 Scheme Part II paragraph 12.
13 Scheme Part II paragraph 2(1).
14 Scheme Part II paragraph 2(3).
15 Scheme Part II paragraph 2(4).

The contract price means *the entire sum payable under the construction contract in respect of the work.*[16] This clearly includes all variations and delay and disruption claims.

The adjudicator will have to assess the submissions of the parties and decide what the cost of the work was. The Scheme provides little help to the adjudicator for this task.

This procedure mirrors very closely what is currently required under most standard construction contracts.

Dates for payment

The Act requires that *every construction contract shall*

(*a*) *provide an adequate mechanism for determining what payments are due under the contract, and when, and*
(*b*) *provide for a final date for payment in relation to any sum which becomes due.*

The parties are free to agree how long the period is to be between the date on which a sum becomes due and the final date for payment.[17]

This is the crux of this part of the Act. The final date for payment is a misnomer. It really has little to do with payment but all to do with the right to suspend performance. However, the Act does imply that an adjudicator has power to order payment to be made if a sum due is not paid by the final date for payment.[17a]

An unpaid party is entitled, if not paid by the final date for payment, to give seven days' notice that, if he is not paid in full within the seven days, he intends to suspend performance.

The Scheme provides criteria for determining when these dates are:

Where the parties to a construction contract fail to provide an adequate mechanism for determining either what payments become due under the contract, or when they become due for payment, or both, the relevant provisions of paragraphs 4–7 shall apply.[18]

16 Scheme Part II paragraph 12.
17 s110(1).
17a s111(4).
18 Scheme Part II paragraph 3.

We have dealt already with the determination of the sums due, so now we must consider when they become due.

Under standard construction contracts generally the mechanism is that the payee submits an account and the payer (and his consultant if there is one) is given a period in which to certify the sum due and pay it. This is normally around 28 days (see Fig. 2).[19]

The Scheme provides only seven days from the end of the *relevant period* or on presentation of claim if that is later (see Fig. 2).

The payment shall *become due on which of the following dates occurs later:*[20]

(*a*) *the expiry of seven days following the relevant period ... , or*
(*b*) *the making of a claim by the payee.*

These periods also apply to contracts which have an adequate mechanism for determining what sums are due but do not define when they become due for payment.[21]

A *claim by the payee*[22] is defined as *a written notice given by the party carrying out work under a construction contract to the other party specifying the amount of any payment or payments which he considers to be due and the basis on which it is, or they are calculated.*

The date when payment is due is the date from which the final date for payment is calculated. The period between these two dates under the Scheme is 17 days (see Fig. 2).

Where the parties to a construction contract fail to provide a final date for payment in relation to any sum which becomes due under a construction contract, the provisions of this paragraph apply.

This final date for the making of any payment ... shall be 17 days from the date that payment becomes due.[23]

The Scheme also curiously defines the amount finally due under a construction contract and when it becomes due. This is quite

19 ICE Conditions of Contract (Sixth Edition) Clause 60(2).
20 Scheme Part II paragraph 4.
21 Scheme Part II paragraph 7.
22 Scheme Part II paragraph 12.
23 Scheme Part II paragraphs 8 (1) & (2).

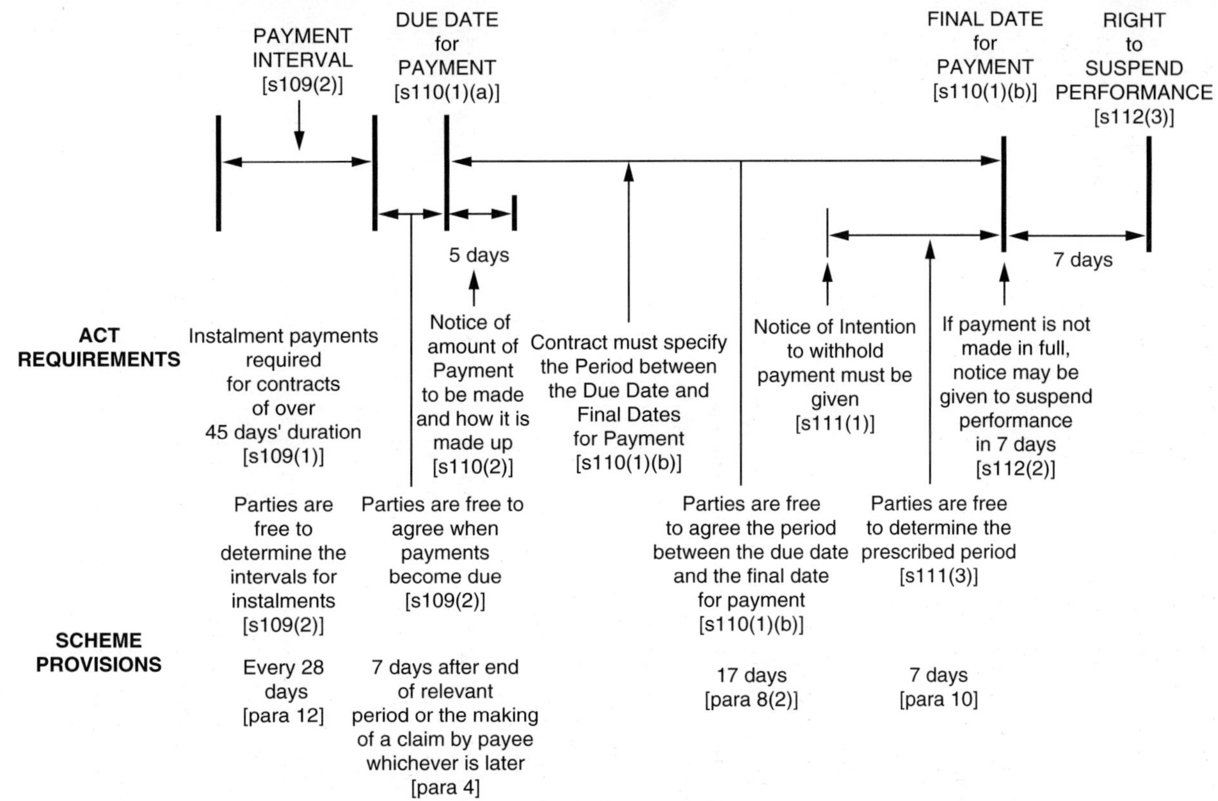

Figure 2. Payment dates and the right to suspend performance

unnecessary since the right to suspend performance is meaningless once the work has been completed!

The Scheme says that the difference between the contract price and the aggregate of the instalments shall be due on a *relevant construction contract* (that is one longer than 45 days) either

(a) 30 days following completion of the work, or
(b) the making of a claim by the payee
whichever is the later.[24]

This provision can cause difficulty if the negotiations for the final certificate are protracted. The problem is what happens if the payee makes more than one claim after 30 days. Negotiations then take place. Agreement is reached on some items which increase the amount agreed by the payer and decrease the amount claimed by the payee. Some money may then be paid on account for the amount accepted by the payer. The payee then reverts to his original position and the payer fails to issue a withholding notice. What is then due? Do the provisions for final date for payment cease since the purpose is to enable work to be suspended or does the sum applied for become payable subject to any abatement?

If the question arises and the payee issues a Notice of Adjudication requiring payment in full, it would be sensible for the payer to immediately issue his own Notice of Adjudication referring all the outstanding matters to the adjudicator.

It may well be because of the operation of Paragraph 8(1) of the Scheme that another adjudicator is appointed but it would be likely that the second adjudicator's Decision will be reached before the first Decision is due for payment.

It would of course be better to avoid the problem altogether if you are the paying party by issuing a withholding notice under s111.

24 Scheme Part II paragraph 5.

For a non-relevant construction contract, i.e. one less than 45 days' duration, the contract sum becomes due on

(*a*) *the expiry of 30 days after the completion of the work, or*
(*b*) *the making of a claim by the payee.*[25]

It is unnecessary to define this any closer because the work will generally be completed before the sum becomes due.

Notice specifying amount of payment

The Act requires that the paying party shall notify the other party at least five days after a payment is due what payment is being made and the basis upon which it is calculated.[26] The Act also requires a paying party, if he is going to withhold any sum due as he may be entitled to do for contra-charges or abatement, to give a notice stating the amount which is to be withheld and the grounds for withholding it.[27]

This again is the normal procedure in much of the construction industry. The contractor puts in his monthly statement and the consultant (architect or engineer) certifies the amount due. These two are not necessarily the same.[28] The certificate normally states how the sum is made up. If the sum to be paid is not the sum certified, again it is common for a notice to be required to say how the sum being paid is calculated.[29]

The Scheme requires that a notice of the value of work done under the contract should be given for every interim payment made (or to be made) regardless of whether any money is actually going to be paid. If there are to be contra charges or set-off then these also require a notice. These two different subjects may be covered by one notice.[30]

25 Scheme Part II paragraph 6.
26 Act s110(2).
27 Act s111(1).
28 See ICE Conditions of Contract (Sixth Edition) Clauses 60(1) & (2).
29 See ICE Conditions of Contract (Sixth Edition) Clause 60(10).
30 Act s111(1) 2nd paragraph.

The Scheme more or less repeats the words of the Act.[31] A *party to a construction contract shall, not later than five days after the date on which any payment:*

(*a*) *becomes due from him, or*
(*b*) *would have become due, if*
  (*i*) *the other party had carried out his obligations under the contract, and*
  (*ii*) *no set-off or abatement was permitted by reference to any sum claimed to be due under one or more other contracts, give notice to the other party to the contract specifying the amount (if any) of the payment he has made or proposes to make, specifying to what the payment relates and the basis on which that amount is calculated.*[32]

This notice, which is generally referred to as 'section 110 notice', is intended to be the valuation of the work done under the contract. The second part (*b*) means that a notice will be due under every construction contract where work has been done during the relevant period or when instalments become due.

The two legal terms of set-off and abatement need some explanation.

Set-off allows cross-claims to be deducted from sums due from one party to the other. The test as to whether or not a cross-claim can be set off is whether it is so closely connected with the claim that it would be manifestly unjust to allow the claim without taking into account the cross-claim. The most common form of cross-claim is when a subcontractor delays the main contractor and the contractor sets off his additional overhead costs against the sum due to the subcontractor. It would be manifestly unjust to allow payment for a delay claim made by a sub-contractor without taking into account any delays that the sub-contractor may have caused the main contractor.

31 s110(2).
32 Scheme Part II paragraph 9.

Abatement is the withholding of money for some alleged breach of contract by the other party which reduces the value of the work done. The measure of the abatement is 'how much less the subject matter ... is worth by reason of the breach'.[33] The most common form of abatement is for defective work.

This section 110 presupposes that the sum stated in the notice will be paid before the *final date for payment*.

If any sum is intended to be withheld after the final date for payment, the payee must give notice to this effect not later than seven days before the *final date for payment*. This may be stated either in the notice advising the amount to be paid,[34] or in a separate notice.[35]

> *A party to a construction contract may not withhold payment after the final date for payment of a sum due under the contract unless he has given an effective notice of intention to withhold payment.*[36]
>
> *To be effective such notice must specify:*
>
> *(a) the amount proposed to be withheld and the ground for withholding payment, or*
>
> *(b) if there are more than one ground, each ground and the amount attributable to it, and must be given not later than the prescribed period before the final date for payment.*[37]

The Scheme stipulates that the prescribed period *is seven days before the final date for payment determined either in accordance with the construction contract, or when no such provision is made in the contract, in accordance with*[38] the provisions for determining the final date for payment in the Scheme (as described above) (see Fig. 2).

This notice is generally referred to as either the 'section 111 notice' or the 'withholding notice'. It should be noted that the notice is for withholding payment from a 'sum due' under the

34 s110(1) second sentence.
35 s111.
36 s111.
37 s111(2).
38 Scheme Part II paragraph 10.

contract. When a dispute arises concerning payment it is normal for the paying party to dispute that the sum applied for was in fact due. The adjudicator will in most cases have to determine the amount due at any given time. The difficulty for the adjudicator is in deciding the depth to which he should investigate to determine the sum due.

Adjudicators may have widely differing views. Some adjudicators may take it upon themselves to value the works in detail and others may take the view that if neither party is prepared to do a detailed evaluation then it is not for the adjudicator to do so.

Caution would therefore suggest that if a party wishes to challenge a statement prepared by a payee he should prepare in some detail, and in a checkable form, his arguments as to why the sum claimed is wrong.

'In considering a dispute, an adjudicator will make his own valuation of the claim before him and in doing so, he may abate the claim in respects not mentioned in the notice of intention to withhold payment. But he ought not to look into abatements outside the four corners of the claim unless mentioned in the notice to withhold payment.'[38a]

The notice to withhold payment must be in writing and cannot be made before the application for payment is made.[38b] A withholding notice will be required before sums can be either set-off or abated against a sum due. Set-offs arising after the due date for payment cannot be considered.[38c] If the work was simply overmeasured then there would be no need for a withholding notice as there would not be an abatement because the sum claimed was never contractually due. If it is claimed that work done was defective then that would be an abatement and a withholding notice would be required.

38a Judge Peter Bowsher in *Whiteways Contractors (Sussex) Limited* v. *Impresa Castelli Construction UK Limited.*

38b *Strathmore Building Services (Birmingham) Ltd* v. *Sindall Ltd.*

38c *Northern Developments Ltd* v. *J & J Nichol.*

The right to suspend performance

The Act says *where a sum due under a construction contract is not paid in full by the final date for payment and no effective notice to withhold payment has been given the person to whom the sum is due has the right (without prejudice to any other right or remedy) to suspend performance of his obligations under the contract to the party by whom payment ought to have been made ('the party in default').*[39]

There is a strict procedure for anyone intending to exercise this right. It is important that it be strictly adhered to because the consequences of being wrong could be disastrous. The suspending party might then find himself to be in breach of his contract and the other would be entitled to damages for any loss that the suspension caused him.

> *The right may not be exercised without first giving to the party in default at least seven days' notice of intention to suspend performance, stating the ground or grounds on which it is intended to suspend performance.*[40]

There are a number of points to note before suspension is contemplated. First, a notice of intention to suspend performance can be given only if no notice to withhold payment has been given.

Secondly, if the notice is considered to be wrong because the sum to be withheld is too large, or the grounds are thought to be wrong, the appropriate action is to refer the matter to the adjudicator.

The Act provides for circumstances where the adjudicator considers that some of the money should not be withheld:

> *Where an effective notice of intention to withhold payment is given but on the matter being referred to adjudication it is decided that the*

39 s112(1).
40 s112(2).

> *whole or part of the amount should be paid, the decision shall be construed as requiring payment not later than:*
>
> *(a) seven days from the date of the decision, or*
> *(b) the date which apart from the notice would have been the final date for payment, whichever is the later.*[41]

The words 'requiring payment' are important because they imply that the adjudicator can order the payment of monies by one party to the other. The adjudicator in his decision will be bound by these time periods.

Normally, the decision of the adjudicator will be given after the final date for payment so that the payee, if not paid seven days after the decision, may issue the notice of intention to suspend performance if he has not been paid the sum decided by the adjudicator.

If the final date for payment has not passed then the normal routine prevails.

It would always be sensible to get the decision of an adjudicator before performance is suspended unless neither of the required notices have been issued.

The right to suspend performance ceases when the party in default makes payment in full of the amount due.[42]

The Act does not compensate the party who suspends performance for the additional costs incurred in demobilising and re-mobilising, or any other costs that may be incurred. It only provides for an extension of time for completion – nothing else.

> *Any period during which performance is suspended in pursuance of the right conferred by this section shall be disregarded in computing for the purposes of any contractual time limit the time taken, by the party exercising the right or by a third party, to complete any work directly or indirectly affected by the exercise of the right.*
>
> *Where the contractual time limit is set by reference to a date rather than a period, the date shall be adjusted accordingly.*[43]

41 s111(4).
42 s112(3).
43 s112(4).

The conclusion must be that only if payment is drastically delayed will it be economically worth suspending performance.

Some standard contracts allow the contractor to recover the additional costs incurred as the result of such suspension.[43a]

Prohibition of conditional payment provisions

The Act rendered 'pay when paid' clauses ineffective.

A *provision making payment under a construction contract conditional on the payer receiving payment from a third person is ineffective, unless that third person, or any other person payment by whom is under the contract (directly or indirectly) a condition of payment by that third person, is insolvent.*[44]

The exception will protect main contractors from having to pay subcontractors if the employer becomes insolvent. The same is true for subsubcontractors and so on down the chain. The policy reason must be to spread the risk associated with insolvency as widely as possible.

The Act defines what it means by insolvency and anyone experiencing the insolvency of another should contact a solicitor for advice. Such advice is beyond the scope of this book.

In relation to 'pay when paid' clauses, the Scheme only applies when there is a 'pay when paid' clause in the contract and the parties have agreed no other conditions. In that case all the provisions of the Scheme (payment by instalments, valuation of instalments, due date for payment and the final date for payment) detailed above will apply.[45]

Summary

The main provisions of the Scheme for payment are:

- For contracts over 45 days there shall be periodic payments every 28 days.

43a See JCT clause 26.2.10; ECC clause 60.7.

44 s113(1).

45 Scheme Part II paragraph 11.

- Period payments are calculated by evaluating the total work done and subtracting previous instalment payments.
- Payments are due on the later of seven days after the end of the relevant period or on presentation of the claim.
- The final date for payment is 17 days after the sum becomes due.
- A notice must be given declaring what sum has been or will be paid within 5 days after the sum becomes due.
- An effective notice withholding any payment must be given at least 7 days before the final date for payment.
- A notice giving seven days notice of suspension of performance may be given if payment in full (subject to deductions contained in a valid withholding notice) is not received by the final date for payment.
- The right to suspend work ceases on payment in full.
- The Scheme provides for an extension of the time for completion only when the works are suspended, but there is no compensation.
- When a 'pay when paid' clause is ineffective and no other terms of payment have been agreed the provisions of the Scheme apply.

13

The Act

The Act is the Housing Grants, Construction and Regeneration Act 1996. The Act is a hybrid and covers a number of different subjects associated with the construction industry. The adjudication and payment provisions all arise out of Part II of the Act.

The Act was introduced to redress the balance between the paid and the paying parties. Cash flow is paramount in an industry which runs on very limited capital in relation to turnover. All too often the paying party has raised overstated counter-claims (or even imaginary ones) in order to reduce the payments that it makes. Such acts have resulted in the creditor company going into liquidation and the paying party avoiding making payment altogether.

Although it is the adjudication provisions which have attracted the widest publicity it is probably the payment provisions which have had the greatest effect. This is because the paying party is now obliged to state how much he intends to pay and why. If there is a difference between the parties on what should be paid this can now be referred immediately to an adjudicator.

However, it is the courts that have been responsible for a large part of the Act's success. Their support and willingness to enforce adjudicators' decisions both speedily and regularly have removed most of the benefits derived from delaying payments. The revolutionary aspect of the Act is the principle that a temporary decision should be enforceable by the courts pending final resolution.

The Housing Grants, Construction and Regeneration Act 1996 consists of three parts. The one we are interested in is Part II, which starts at s104. At the time of writing there is no proposal for amending the Act although it has not been ruled out. Amendments have been made to the Scheme.

This book includes reference to the judgments of the court as well as observations that have been made by various commentators, both lawyers and practitioners. At the time of writing there had been four judgements of the Court of Appeal, but over 200 in the courts of first instance, generally, in enforcement proceedings.

A construction contract [section 104]

The Act applies to construction contracts.

Part II Construction contract

Introductory provisions

104 Construction contracts

(1) *In this Part a 'construction contract' means an agreement with a person for any of the following—*
   - (*a*) *the carrying out of construction operations;*
   - (*b*) *arranging for the carrying out of construction operations by others, whether under sub-contract to him or otherwise;*
   - (*c*) *providing his own labour, or the labour of others, for the carrying out of construction operations.*

The Act only applies to 'construction contracts'. A construction contract is one for 'construction operations' as defined in the Act. Section 105 defines what is meant by a construction operation.

(2) *References in this Part to a construction contract include an agreement—*
   - (*a*) *to do architectural, design, or surveying work, or*
   - (*b*) *to provide advice on building, engineering, interior or exterior decoration or on the laying-out of landscape,*

   in relation to construction operations.

It can be seen that this covers the work of architects, engineers, contractors, subcontractors and all who give advice on construction operations. It has been held that quantity surveying work in respect of claims for additional payment is not advice in relation to construction operations and therefore is not covered by the Act.[1]

It has been questioned as to whether legal advice in relation to the drafting of the contract documents is advice in relation to construction operations but it is generally considered that it is not.

(3) *References in this Part to a construction contract do not include a contract of employment (within the meaning of the Employment Rights Act 1996) (1996 c. 18).*

The Act cannot be used for the settlement of disputes between an employer and his employees. Under the Employment Protection (Consolidation Act) 1978 such disputes are dealt with by Employment Tribunals.

(4) *The Secretary of State may by order add to, amend or repeal any of the provisions of subsection (1), (2) or (3) as to the agreements which are construction contracts for the purposes of this Part or are to be taken or not to be taken as included in references to such contracts.*

No such order shall be made unless a draft of it has been laid before and approved by a resolution of each House of Parliament.

The Secretary of State is given power to amend what is a construction contract and therefore the range or type of contract to which the Act will apply. A similar power is given to the Secretary of State in s105 to vary what is considered to be a construction operation. However, since the Act is directed at the construction industry it would not be possible for the Secretary of State to widen that application of the Act to contracts which have nothing to do with construction.

(5) *Where an agreement relates to construction operations and other matters, this Part applies to it only so far as it relates to construction operations.*

1 *Fence Gate Ltd* v. *James R. Knowles Limited* [TCC 31 May 2001].

This section may have far-reaching effects. It extends the requirements of the Act into all agreements which contain construction operations. For instance, where a lease for a property contains a clause that the landlord may on completion of a tenancy repair any defects if the tenant fails to do them the Act will apply to the arrangements for making good those deficiencies.

An agreement relates to construction operations so far as it makes provision of any kind within subsection (1) or (2).

(6) *This Part applies only to construction contracts which—*
  (*a*) *are entered into after the commencement of this Part, and*
  (*b*) *relate to the carrying out of construction operations in England, Wales or Scotland.*

The Act commenced on 1 May 1998, and applies to all construction contracts entered into after that date. They may have been started before that date but if the parties signed a formal agreement after the date then all the operations carried out under that contract will be subject to the Act.[2]

Only construction contracts in England, Wales and Scotland are covered by the Act, but similar provisions are in force in Northern Ireland.[3] However, for the purpose of the Act the country boundaries occur at the coastal low water line. This stems from the definition of a construction operation in s105(1).

(7) *This Part applies whether or not the law of England and Wales or Scotland is otherwise the applicable law in relation to the contract.*

Under English law a party is entitled to select the rules of law by which the contract is to be interpreted. There is no problem when both the parties are domiciled in the country where the construction operations are performed but increasingly overseas contractors are being employed for specialist work. In the absence of any provision

2 *Christiani and Nielson* v. *The Lowry*.

3 S.I. 1997/274 (NI) and the Scheme for Construction Contracts in Northern Ireland Regulations (Northern Ireland) 1999 came into force on 1 June 1999.

to the contrary the law of the contract is the law of the country where the contract is formed. Thus if a German main contractor makes a contract with a German subcontractor in Germany for works in England the contract may well be interpreted under German law but that will depend upon wording of the subcontract agreement.

The meaning of construction operations [section 105]

The Act only provides a right to adjudication or the benefits of the payment provisions if the contract is one to which the Act applies. It only applies to construction operations which fall within s105(1), unless the contract itself extends these provisions to other operations. When the contract provisions comply with the Act then an adjudicator will have has jurisdiction over all disputes arising under the contract unless it is restricted in the contract to the activities covered by the Act. The JCT and ICE Contracts cover all activities but the IChemE Conditions, for instance, limit adjudication to the operations covered by the Act.

This section is the cause of many of the jurisdictional challenges that are made to try to stop an adjudication.

105 Meaning of 'construction operations'

(1) *In this Part 'construction operations' means, subject as follows, operations of any of the following descriptions—*

   (*a*) *construction, alteration, repair, maintenance, extension, demolition or dismantling of buildings, or structures forming, or to form, part of the land (whether permanent or not);*

These definitions of construction operations were derived from the s567 of the Income and Corporation Taxes Act 1988. They were, therefore, drafted for a different purpose.

105(1)(a) is a general description of building operations. It is fairly straightforward except for the words 'forming, or to form, part of the land (whether permanent or not)'. These are derived from the law relating to landlord and tenant and most of the case law concerns fittings and fixtures, which the tenant is entitled to take upon completion of a tenancy. In general, things which are the tenant's upon completion of the lease are not 'part of the land'. It has been held that the construction of steel modules for an oil rig, made in a yard at Teeside, were not construction operations because they were not intended to form part of the land at Teeside.[4] It was also held there that certain shop fittings were not construction operations for a similar reason.[5]

(*b*) *construction, alteration, repair, maintenance, extension, demolition or dismantling of any works forming, or to form, part of the land, including (without prejudice to the foregoing) walls, roadworks, power-lines, telecommunication apparatus, aircraft runways, docks and harbours, railways, inland waterways, pipe-lines, reservoirs, water-mains, wells, sewers, industrial plant and installations for purposes of land drainage, coast protection or defence;*

This definition covers civil engineering works. However, the Act only applies to operations that are above low water mark, so sea-platforms and the like will be outwith the scope of the Act. It is suggested that jetties and the like which are founded partly above and partly below low water line will caught by the Act, but any which are wholly below such as dolphins will not be covered.

(*c*) *installation in any building or structure of fittings forming part of the land, including (without prejudice to the foregoing) systems of heating, lighting, air-conditioning, ventilation, power supply, drainage, sanitation, water supply or fire protection, or security or communications systems;*

(*d*) *external or internal cleaning of buildings and structures, so far as carried out in the course of their construction, alteration, repair, extension or restoration;*

It is clear that all fitting out operations fall within the scope of the Act. It was held that an annual maintenance contract for heating

4 *Staveley Industries plc* v. *Odebrecht Oil & Gas Services Ltd* (unreported).

5 *Gibson Lea Retail Interiors Ltd* v. *Marko Self-Services Wholesalers* [24 July 2001].

installations entered into by a local authority fell within the scope of the Act.[6]

(*e*) *operations which form an integral part of, or are preparatory to, or are for rendering complete, such operations as are previously described in this subsection, including site clearance, earth-moving, excavation, tunnelling and boring, laying of foundations, erection, maintenance or dismantling of scaffolding, site restoration, landscaping and the provision of roadways and other access works;*

It would seem that such operations as a preparatory land survey, site investigation (boreholes and trial pits), test piles and load tests would all be subject to the Act.

(*f*) *painting or decorating the internal or external surfaces of any building or structure.*

It should be remembered that maintenance is one of the general activities mentioned in subsubsections (a) and (b) and this subsubsection will also apply to maintenance contracts.

Exceptions

(2) *The following operations are not construction operations within the meaning of this Part—*

(*a*) *drilling for, or extraction of, oil or natural gas;*

(*b*) *extraction (whether by underground or surface working) of minerals; tunnelling or boring, or construction of underground works, for this purpose;*

The first exceptions are clear. Quarrying and mining are not included and associated operations such as sinking shafts, access tunnels, adits and drifts constructed to enable minerals to be extracted are also exempt. This will mean that such operations as

---

6 *Nottingham Community Housing Association Limited* v. *Powerminster Limited* (www.adjudication.co.uk).

coal mining, extracting masonry stones (like Bath Stone), and the digging for aggregates are not covered by the Act.

This will probably exclude the digging for borrow on motorways and the excavation of other fill materials. However, in practice this is not likely to cause a problem since most major contracts will include Act compliant clauses, but it may be a problem for small sub-contractors, if the main contractor wishes to avoid the application of the Act to these works.

> (c) *assembly, installation or demolition of plant or machinery, or erection or demolition of steelwork for the purposes of supporting or providing access to plant or machinery, on a site where the primary activity is—*
>
> *(i) nuclear processing, power generation, or water or effluent treatment, or*
>
> *(ii) the production, transmission, processing or bulk storage (other than warehousing) of chemicals, pharmaceuticals, oil, gas, steel or food and drink;*

This exemption was introduced as a result of submissions made by the process industries, who claimed that they had few disputes and opted out of the Latham enquiry.

'... *on a site where the primary activity is* ... '. This is to be determined on the facts and not necessarily by the description of the site as defined in the construction contract. For a contract for the installation of a generator to enable printing works to function should the Millennium Bug affect the power system it was held that the primary activity on the site was printing and therefore the installation of the generator came within the Act.[7]

The exemption is of limited scope and all contracts in respect of the construction and maintenance of process plants are likely to have operations which are covered by the Act. For instance, all building work, access scaffolding used for erecting the plant,[8] will be

7 *ABB Zantingh Limited* v. *Zedal Building Services Limited* (www.adjudication.co.uk).

8 *Palmers Limited* v. *ABB Power Construction Limited* [1999] BLR 426.

within its scope. However, pipework and electric cables connecting the items of plant,[9] insulation and cladding to pipework, boilers and the like,[10] and the installation, labelling, termination and testing of all field wiring (including the supply of metal containment systems and secondary steel support) will not be covered by the Act.[11]

It would seem that the courts will try to include an operation rather than exclude it from the operation of the Act.

(d) *manufacture or delivery to site of—*
  (i) *building or engineering components or equipment,*
  (ii) *materials, plant or machinery, or*
  (iii) *components for systems of heating, lighting, air conditioning, ventilation, power supply, drainage, sanitation, water supply or fire protection, or for security or communications systems,*

except under a contract which also provides for their installation;

The supply of materials and delivery of all building materials and components is excluded from the operation of the Act unless the contract also includes the supply of those items.

It follows that a dispute as to whether or not an item complies with its specification can be adjudicated under the installation contract but not under the supply contract.

(e) *the making, installation and repair of artistic works, being sculptures, murals and other works which are wholly artistic in nature.*

It would seem that the actual work of art is exempt but not the plinth or foundation upon which it is mounted.

(3) *The Secretary of State may by order add to, amend or repeal any of the provisions of subsection (1) or (2) as to the operations and work to be treated as construction operations for the purposes of this Part.*

---

9 *Homer Burgess Limited* v. *Chirax(Annan) Limited* [2000] BLR 124.

10 *ABB Power Construction Limited* v. *Norwest Holst Engineering Limited* (www.adjudication.co.uk).

11 *ABB Zantingh Limited* v. *Zedal Building Services Limited* (www.adjudication.co.uk).

(4) *No such order shall be made unless a draft of it has been laid before and approved by a resolution of each House of Parliament.*

This definition like the previous definitions for a construction contract, and a construction operation can be amended by the Secretary of State but only after it has been approved by resolution of each House of Parliament.

It is to be hoped that 105(2)(c) will be removed in due course. It only produces a two level dispute procedure for those contracts to which it applies.

Provisions not applicable to contract with residential occupier [section 106]

Again this exemption is restricted. It only applies to a dwelling house or flat which one of the parties occupies or intends to occupy as his residence. Therefore if a developer wishes to occupy a flat in a block of flats that he is developing the exemption possibly would apply to that one flat. However, it has been held that where a residential occupier anticipated occupying 65% of the works as his private residence that was insufficient to avoid the application of the Act.[12]

It would seem therefore that the courts take the view that in a hybrid contract where a party is to be a residential occupier of part of the premises that part has to be a major part of the works carried out.

106 Provisions not applicable to contract with residential occupier

(1) *This Part does not apply*

(*a*) *to a construction contract with a residential occupier (see below), or*

(*b*) *to any other description of construction contract excluded by the operation of this Part by order of the Secretary of State.*

The reference in (b) to an order by the Secretary of State will only apply to orders in connection with contracts with residential occupiers.

12 *Samuel Thomas Contruction* v. *Blick & Blick* (www.adjudication.co.uk).

Who is a residential occupier?

(2) *A construction contract with a residential occupier means a construction contract which principally relates to operations on a dwelling which one of the parties to the contract occupies, or intends to occupy, as his residence.*

In this subsection 'dwelling' means a dwelling-house or a flat; and for this purpose—
'dwelling-house' does not include a building containing a flat; and
'flat' means separate and self-contained premises constructed or adapted for use for residential purposes and forming part of a building from some other part of which the premises are divided horizontally.

The intention is clear but whether a maisonette is covered could be arguable.

(3) *The Secretary of State may by order amend subsection (2).*
(4) *No order under this section shall be made unless a draft of it has been laid before and approved by a resolution of each House of Parliament.*

So be it.

Agreement to be in writing [section 107]

This whole section is a reproduction of Section 5 of the Arbitration Act 1996.

Nominally to be affected by the Act the construction contract has to be in writing but ...

107 Provisions applicable only to agreements in writing

(1) *The provisions of this Part apply only where the construction contract is in writing, and any other agreement between the parties as to any matter is effective for the purposes of this Part only if in writing.*

The expressions 'agreement', 'agree' and 'agreed' shall be construed accordingly.

This is clear but the remainder of the section shows it to be totally misleading because the definition of 'writing' is extremely wide indeed.

(2) *There is an agreement in writing—*

(*a*) *if the agreement is made in writing (whether or not it is signed by the parties),*

(*b*) *if the agreement is made by exchange of communications in writing, or*

(*c*) *if the agreement is evidenced in writing.*

These three items are further explained in the next four subsections, but it should be noted that under subsection (6) it says that references ... *to anything being written or in writing include its being recorded by any means.* This will include a tape recording of a conversation. Thus

(3) *Where parties agree otherwise than in writing by reference to terms which are in writing, they make an agreement in writing.*

A telephone conversation agreeing that certain works are to be carried out in accordance with a previous contract would constitute an agreement in writing. The difficulty with this, as with any contract agreed orally, is proving the terms.

What this subsection does is to make such an oral agreement a 'construction contract' within the meaning of the Act, and that gives an adjudicator jurisdiction to decide what its terms are.

(4) *An agreement is evidenced in writing if an agreement made otherwise than in writing is recorded by one of the parties, or by a third party, with the authority of the parties to the agreement.*

This is the situation referred to in (2)(b) above. Thus minutes of meetings at which agreement is reached for certain works to be undertaken would constitute a construction contract. Indeed a tape recording of a telephone conversation between the parties at which agreement was reached, which was recorded by one of them (it would seem even without the agreement or knowledge of the other party) would constitute a contract in writing.

It is necessary for the terms of the contract to be in writing, or at least the relevant ones for the Act to apply. It was held that evidence of a contract such as invoices, payment certificates, minutes of meetings were insufficient to comply with this subsection. There had to be evidence of the terms of the contract. 'What is important is that the terms of the agreement material to the issue or issues giving rise to the reference should be clearly recorded in writing, not that every term, however trivial or unrelated to those issues, should be expressly recorded or incorporated by reference'.[13]

(5) *An exchange of written submissions in adjudication proceedings, or in arbitral or legal proceedings in which the existence of an agreement otherwise than in writing is alleged by one party against another party and not denied by the other party in his response constitutes as between those parties an agreement in writing to the effect alleged.*

This subsection has caused some difficulty.

The first difficulty has been the words ... *to the effect alleged* in the last line.

Under the Arbitration Act 1996 it is the agreement to arbitrate that has to be in writing. This is a single issue matter. That is not the case in this Act where all the terms of the contract may be at issue and it cannot be right that if one party does not deny that there is an oral contract in adjudication of arbitral or legal proceedings then the terms of the contract are those alleged by the referring party. It has been held that this subsection will only have that effect if there has been a previous adjudication (arbitration or litigation) in which the terms alleged were not denied.[14]

The effect of that ruling is that the parties can agree that there was an oral contract provided that they set out the terms that have been agreed and can then have their dispute adjudicated.

However, there is a hidden danger. If the dispute is, for instance, a dispute as to whether certain work complies with the specification

---

13 Lord Justice Auld: *RTJ Consulting Engineers Ltd* v. DM *(Northern Ireland) Ltd* (CA) GWS.

14 *Grovedeck* v. *Capital Demolition Limited* [2000] BLR 181.

and both parties agree that the best solution is to refer the question to an appropriate technical adjudicator, they will have by so doing agreed that they have a construction contract. If a second dispute then arises concerning payment, in the absence of complying payment terms in their oral contract, they may have imported the payment terms in the Scheme into their contract by agreeing to the first adjudication.

(6) *References in this Part to anything being written or in writing include its being recorded by any means.*

This, as is previously pointed out, will include a tape recording, but will also include fax, telex, and e-mail.

Adjudication [section 108]

This Section has been the subject of detailed analysis elsewhere in this book and more detailed comments may be found in the individual chapters. It is only the intention here to identify the general points.

108

(1) *A party to a construction contract has the right to refer a dispute arising under the contract for adjudication under a procedure complying with this section.*

For this purpose 'dispute' includes any difference.

The Act therefore gives a party to a construction contract a 'right' to adjudication. It does not negate any other form of dispute resolution the parties may agree to, but supplements them should any party so wish.

'dispute includes any difference' The reason for this curious phrase is that in *Sykes v. Fine Fare* it was considered by Lord Justice Danckwerts that a difference could cover a wider situation than a dispute.

There must be a dispute between the parties for the adjudicator to have jurisdiction and a dispute means whatever claims, heads of claim, issues, contentions or causes of action that were in dispute at the moment that the referring party first intimated an adjudication reference and chose to crystallize into an adjudication reference.[15] However, a Scottish court hinted that this definition might be too wide[16] and in a later case it was held that the dispute referred must be the issues as they were argued at the time of the issue of the Notice of adjudication and should not include arguments that had not been rehearsed between the parties.[17]

The dispute must arise 'under the Contract'. This is restrictive and removes a number of disputes from the jurisdiction of an adjudicator. Tortious claims for negligence, trespass or nuisance are outwith his jurisdiction, and so are disputes concerning settlement agreements.

The 'right' bestowed by subsection (1) is satisfied by one of two procedures:

(*a*) Either the contract contains the necessary provisions, or
(*b*) The Scheme for Construction Contracts applies.[18]

Subsections (2)–(5) set out what are referred to as the 'eight compliance points'. These are the eight provisions that the contract adjudication clause must contain to comply with Part II of the Act. They may be added to but not in such a way as to negate any of these eight principles.

(2) *The contract shall—*

 (*a*) *enable a party to give notice at any time of his intention to refer a dispute to adjudication;*

15 *Fastrack Contractors Limited* v. *Morrison Construction Ltd and Impregilo UK Ltd* [2000] BLR 168.

16 *Barr Limited* v. *Law Mining Limited* [15 June 2001 SWS].

17 *Edmund Nuttall Ltd* v. *R.G. Carter Limited* [21 March, 2002] (Bailii).

18 Strictly The Scheme for Construction Contracts Regulations (England and Wales) 1998 No. 649. There are separate Schemes for Scotland and Northern Ireland, but they are all very similar.

(b) *provide a timetable with the object of securing the appointment of the adjudicator and referral of the dispute to him within 7 days of such notice;*
(c) *require the adjudicator to reach a decision within 28 days of referral or such longer period as is agreed by the parties after the dispute has been referred;*
(d) *allow the adjudicator to extend the period of 28 days by up to 14 days, with the consent of the party by whom the dispute was referred;*
(e) *impose a duty on the adjudicator to act impartially; and*
(f) *enable the adjudicator to take the initiative in ascertaining the facts and the law.*

These are the first six compliance points. They are clear in both the drafting and intent.

Most standard construction contracts include these provisions word for word to make sure the contract complies.

There is some controversy as to whether the 'notice of dissatisfaction' procedure built into the ICE and IChemE forms of contract actually complies with the Act because it delays the time a party may apply for adjudication. It is argued that it offends (2)(a) because the notice of intention cannot be given at any time. However, it probably matters very little. It is a sensible procedure because it enables parties to clarify exactly what they are arguing about before they bring in a third party to solve their differences. If a party is that desperate for a quick decision he can always apply for adjudication under the Scheme and most adjudicators would proceed on the basis of the Scheme.

(3) *The contract shall provide that the decision of the adjudicator is binding until the dispute is finally determined by legal proceedings, by arbitration (if the contract provides for arbitration or the parties otherwise agree to arbitration) or by agreement.*

The parties may agree to accept the decision of the adjudicator as finally determining the dispute.

This is the seventh compliance point. It is the kernel of the Section. It is the subsection which gives the adjudicator's decision its 'temporarily binding' effect.

At the time the Act was drafted it was not known how this provision would be implemented if a party chose to ignore what the adjudicator decided. It was decided in the first case to arise[19] that the courts would enforce the decision summarily. The Technology and Construction Courts have streamlined their procedures and these applications are now normally dealt with within two weeks of application.

It should also be noted that it is the dispute that is referred to arbitration or litigation if one or more of the parties disagree with the adjudicator's Decision. It is not the Decision that is appealed.

It is not clear what the second sentence means. The problem is when may they so decide? In the contract, after the dispute has been referred or after the decision has been made? If the first, then it has been suggested that the procedure would then be arbitration and subject to the Arbitration Act 1996. If the last applies then the sentence adds very little to the words *or by agreement* in the previous sentence. We shall have to await developments.

(4) *The contract shall also provide that the adjudicator is not liable for anything done or omitted in the discharge or purported discharge of his functions as adjudicator unless the act or omission is in bad faith, and that any employee or agent of the adjudicator is similarly protected from liability.*

This is the eighth and final compliance point.

It is an indemnity from the parties that they will be unable to sue the adjudicator for his conduct of the adjudication unless he acts *in bad faith*. It is only a limited indemnity because it does not prevent third parties from suing the adjudicator because they are not parties to the contract containing the indemnity. Bad faith is not a concept

19 *Macob Civil Engineering* v. *Morrison Construction* (1999) 3BLR 93.

clearly defined in English law and is likely to be difficult to prove. Clearly, if the adjudicator accepted a bribe from one of the parties he would be acting in bad faith but negligent conduct of the hearing without any intention of benefiting one or other of the parties is unlikely to be considered bad faith.

That concludes the requirements for a compliant contract provision.

The scheme's application

(5) *If the contract does not comply with the requirements of subsections (1) to (4), the adjudication provisions of the Scheme for Construction Contracts apply.*

This gives the alternative to a compliant contract provision.

Enforcement of the decision

Surprisingly, how an adjudicator's decision was to be enforced was unclear when the Act was passed. If the Scheme was to apply, and it applied as implied terms of the contract,[20] some form of enforcement was required and the draughtsmen of the Act envisaged using the Arbitration Act. This they did by introducing a provision relying upon s 42 of the Arbitration Act 1996. There is no reported case of this procedure ever having been used in England or Wales.

The provision for Scotland has been used but only to the extent that it brings enforcement within the normal court procedures.

(6) *For England and Wales, the Scheme may apply the provisions of the Arbitration Act 1996 with such adaptations and modifications as appear to the Minister making the scheme to be appropriate.*

For Scotland, the Scheme may include provision conferring powers on courts in relation to adjudication and provision relating to the enforcement of the adjudicator's decision.

20 s114 (4).

Payment

The payment provisions have probably had a greater impact on the conduct of the construction industry than adjudication has, although adjudication has been the means of enforcing them.

Cash flow, Lord Denning said, was the life blood of the construction industry. So it is, and the payment provisions were introduced to redress the balance between paying and paid parties to a contract.

Besides setting out time for payment, the provisions are intended to require a creditor party to set out in detail what he is owed and a paying party to say what he is paying and why. These statements if not agreed can then be referred to an adjudicator and an enforceable decision on what should be paid obtained quickly.

This is the real novelty of the Act. Prior to its coming into effect if a dispute arose as to what should be paid to a creditor no money generally passed between the parties until the matter was finally settled by the highest court achievable. This allowed a practice to develop of the paying party putting forward exaggerated or even spurious claims to set off against sums agreed to be due to the creditor party. The Act put a stop to such procedures because the adjudicator can now decide the amount that can be withheld.

The Act provides:

(*a*) Payment by instalments for contracts over 45 days duration [s109].
(*b*) A due date for the payment.
(*c*) A final date for payment.
(*d*) Notice, within 5 days of the due date for payment, of the amount to be paid and how it is calculated [s110].
(*e*) Notice if the paying party is going to withhold sums from the sum due [s111].
(*f*) Right to suspend performance after 7 days notice if payment is not received by the final date for payment.
(*g*) 'Pay when paid' clauses are ineffective except if the paying party is in liquidation.

The general provisions are covered by s109.

109 Entitlement to stage payments

(1) *A party to a construction contract is entitled to payment by instalments, stage payments or other periodic payments for any work under the contract unless—*

(*a*) *it is specified in the contract that the duration of the work is to be less than 45 days, or*

(*b*) *it is agreed between the parties that the duration of the work is estimated to be less than 45 days.*

The intention is clearly that the provisions apply to contracts with a duration of more than 45 days. Whilst the subsection requires either the contract to specify the duration or the parties to agree that it is to be over 45 days, it is suggested that the courts will be slow to say the provisions do not apply because neither of these factors are to be found expressly in the agreement between the parties. The court is likely to determine the parties' intentions from an examination of the requirements of the work.

(2) *The parties are free to agree the amounts of the payments and the intervals at which, or circumstances in which, they become due.*

The policy behind this section is that the parties are free to agree their payment terms but they must be clearly stated in the contract. There is nothing to stop the parties agreeing that there will be only one instalment to be paid upon completion of the works provided that that is expressly stated in the contract. The aim is that both parties will know precisely what payment is due and when it to be paid.

(3) *In the absence of such agreement, the relevant provisions of the Scheme for Construction Contracts apply.*

The payment provisions apply individually in contrast with the adjudication provisions which are applied in total if the contract does not comply.

In subsection (1) the individual components are:

(*a*) The payment interval.

(*b*) How the payments are to be calculated, and

(c) The due date for payment.

The Scheme defines these as:

(*a*) 28 days,
(*b*) the basis of costs to date, and
(*c*) 7 days after the end of the relevant interval or on the making of a claim by the payee, whichever is the later.

(4) *References in the following sections to a payment under the contract include a payment by virtue of this section.*

This provision brings these payments into the procedures for payment detailed in the later sections.

Dates for payment [s110]
The amount to be paid and dates when it will be paid are all important for cash flow. The Act requires the amount and appropriate dates to be determinable with certainty. Clauses where payment is determined by certification by an independent third party are likely to fail this test unless there is a maximum period within which certification is to be made after the end of the payment interval or presentation of an invoice.

110 Dates for payment

(1) *Every construction contract shall—*
(*a*) *provide an adequate mechanism for determining what payments become due under the contract, and when, and*
(*b*) *provide for a final date for payment in relation to any sum which becomes due.*

The parties are free to agree how long the period is to be between the date on which a sum becomes due and the final date for payment.

In most construction contracts the method of assessing the amount due is determined either by a bill of quantities or an activity schedule. These may be either based upon an invoice from the payee or certification on behalf of the payer. All of these would provide the basis for an adequate mechanism for determining the 'sum due'.

The contract must also provide an adequate mechanism for determining when that sum becomes due. There are a number of alternatives in common usage in the construction industry. A number of days after:

(*a*) Presentation of an invoice by the payee.
(*b*) Certification on behalf of the payer.
(*c*) Finite dates for lump sum payments have been reached.

There must also be a final date for payment. It was the intention of the Act that this date would be the trigger to the right to suspend performance if full payment was not made by this date. However, it has become clear that the industry has accepted that this will be the date when payment will be made.

Notice of the amount due

The Act requires the paying party to state the amount that is due under the contract. This is not necessarily the sum that will be paid. The amount due under the contract may be reduced for contra-claims and set offs.

(2) *Every construction contract shall provide for the giving of notice by a party not later than five days after the date on which a payment becomes due from him under the contract, or would have become due if—*

(*a*) *the other party had carried out his obligations under the contract, and*

(*b*) *no set-off or abatement was permitted by reference to any sum claimed to be due under one or more other contracts,*

specifying the amount (if any) of the payment made or proposed to be made, and the basis on which that amount was calculated.

The notice required under this section is known as the 'section 110 notice'. It is given by the payer within 5 days of the due date. The person giving the notice is the payer, that will be the Employer under a main contract. There is no reason why this notice should not be given on his behalf by his agent — the architect or engineer.

The wording of this subsection is not necessarily the easiest to understand. It combines two very different principles — The value of the works executed and the amounts of any deductions that the payer claims to be due to him from the payee.

A set off is the amount of a cross claim (or counter-claim) that the payer is allowed to deduct from sum due to be paid by him. This entitlement may arise from a number of principles:

(*a*) An arguable cross-claim whose amount is known or can be ascertained with certainty. These would be matters under the subject contract.
(*b*) Equitable set off: Claims which are so closely connected with the claim for payment that it would be manifestly unjust to allow the claim without taking into account the cross-claim.
(*c*) Contractual set off: This when there is a term in the contract which expressly permits set off of certain cross-claims.
(*d*) Statutory set off: A statute, for instance, the Insolvency Act 1986, expressly permits a set off.

A set off is a defence and 'can be used as a shield not as a sword',[21] which means that it can be used to reduce the claim but not for a reverse payment.

Abatement is a reduction from the amount claimed resulting from a breach of contract which reduces the value of the works carried out. Thus if work for which payment is claimed is considered to be defective or not in accordance with the specification, or has been overmeasured the value of the defective work or overmeasure may be deducted as a right and if that amounts to a reverse payment so be it.

The section 110 notice therefore values the work carried out including any abatement as a first step. It states how this valuation was arrived at. This must always be done. If the sum to be paid is the valuation the second stage is not necessary.

As a second step it then identifies the amount to be paid which may include any set off arising from the subject contract and any

21 Cockburn CJ in *Stoke* v. *Taylor* (1880) 5QBD 575.

other contract for which set off is permitted. The payer has a second opportunity to state any deduction from the amount due that he may wish to make in the withholding notice issued under s111.

All the amounts to be deducted should state the grounds for their deduction as this is a requirement of s111.

(3) *If or to the extent that a contract does not contain such provision as is mentioned in subsection (1) or (2), the relevant provisions of the Scheme for Construction Contracts apply.*

This subsection therefore requires the contract to provide:

(*a*) An adequate mechanism for determining the sum due.

(*b*) An adequate mechanism for determining when that sum is due.

(c) A final date for payment.

The Scheme provides:

(*a*) The mechanism for determining the sum due consists of:

a. The value of any work performed in the relevant period.

b. An amount for materials manufactured or on site if the contract provides for their payment.

c. Such other amounts as the contract provides should be paid.

The withholding notice

If payment of the full amount due is not to be paid and the Section 110 notice does not detail any deductions to be made the payer is required to issue a notice to that effect prior to the final date for payment.

111 Notice of intention to withhold payment

(1) *A party to a construction contract may not withhold payment after the final date for payment of a sum due under the contract unless he has given an effective notice of intention to withhold payment.*

The notice mentioned in section 110(2) may suffice as a notice of intention to withhold payment if it complies with the requirements of this section.

This notice is generally referred to as the 'section 111 notice' or 'the withholding notice'.

It is only required if the section 110 notice does not comply with the requirement of the section.

(2) *To be effective such a notice must specify—*

(*a*) *the amount proposed to be withheld and the ground for withholding payment, or*

(*b*) *if there is more than one ground, each ground and the amount attributable to it,*

and must be given not later than the prescribed period before the final date for payment.

The grounds do not need to be valid grounds under the contract for the payer to be entitled to make the deduction. He just has to state his reason. If the payee then considers that ground to be invalid a dispute will arise which can then be referred to adjudication under s108 (see subsection 4 below). If the adjudicator considers a ground to be invalid he will direct that the sum deducted be paid.

(3) *The parties are free to agree what that prescribed period is to be.*

In the absence of such agreement, the period shall be that provided by the Scheme for Construction Contracts.

The period under the Scheme is 7 days before the final date for payment.

(4) *Where an effective notice of intention to withhold payment is given, but on the matter being referred to adjudication it is decided that the whole or part of the amount should be paid, the decision shall be construed as requiring payment not later than—*

(*a*) *seven days from the date of the decision, or*

(*b*) *the date which apart from the notice would have been the final date for payment,*

whichever is the later.

This completes the circle. If the ground for withholding is not a valid ground the matter is referred to the adjudicator who can order the withheld sum to be paid and the courts will enforce the decision if the sum is not paid within 7 days. The opportunities for wrongly withholding payment are thus almost nil.

Suspension of performance [s112]

Under English law lack of payment is considered a breach of contract which does not permit the other party to withhold performance. In construction contracts this can be very damaging because construction is a high turnover low margin business.

Section 112 permits an unpaid party to suspend performance.

112 Right to suspend performance for non-payment

(1) *Where a sum due under a construction contract is not paid in full by the final date for payment and no notice to withhold payment has been given, the person to whom the sum is due has the right (without prejudice to any other right or remedy) to suspend performance of his obligations under the contract to the party by whom payment ought to have been made ('the party in default').*

(2) *The right may not be exercised without first giving to the party in default at least seven days' notice of intention to suspend performance stating the ground or grounds on which it is intended to suspend.*

(3) *The right to suspend performance ceases when the party in default makes payment in full of the amount due.*

(4) *Any period during which performance is suspended in pursuance of the right conferred by this section shall be disregarded in computing for the purposes of any contractual time limit the time taken, by the party exercising the right or by a third party, to complete any work directly or indirectly affected by the exercise of the right.*

Where the contractual time limit is set by reference to a date rather than a period, the date shall be adjusted accordingly.

It should be noted that the right given is to suspend performance and that will include all the services being rendered including site security, insurance, pumping and dewatering, protection from weather and any other contractual requirement.

The period of suspension is taken into account when assessing liability of delay damages but there is no compensation for the additional costs incurred by the payee as a consequence of the suspension. However, many of the standard contracts have covered these costs.

Prohibition of conditional payment provisions [s113]

Pay when paid clauses have always been disliked by subcontractors because it has always been very difficult to find out whether payment has been made by the ultimate payer and if so how much in relation to the subcontract works.

113 Prohibition of conditional payment provisions

(1) A *provision making payment under a construction contract conditional on the payer receiving payment from a third person is ineffective, unless that third person or any other person payment by whom is under the contract (directly or indirectly) a condition of payment by that third person, is insolvent.*

The sole exception for allowing such clauses is if the ultimate payer becomes insolvent. This is so that the first supplier does not bear all the risk of a client going into liquidation.

(2) *For the purposes of this section a company becomes insolvent—*
- (*a*) *on the making of an administration order against it under Part II of the insolvency Act 1986,*
- (*b*) *on the appointment of an administrative receiver or a receiver or manager of its property under Chapter 1 of Part 111 of that Act, or the appointment of a receiver under Chapter 11 of that Part,*
- (*c*) *on the passing of a resolution for voluntary winding-up without a declaration of solvency under section 89 of that Act, or*
- (*d*) *on the making of a winding-up order under Part IV or V of that Act.*

(3) *For the purposes of this section a partnership becomes insolvent—*
   (*a*) *on the making of a winding-up order against it under any provision of the Insolvency Act 1986 as applied by an order under section 420 of that Act, or*
   (*b*) *when sequestration is awarded on the estate of the partnership under section 12 of the Bankruptcy (Scotland) Act 1985 or the partnership grants a trust deed for its creditors.*

(4) *For the purposes of this section an individual becomes insolvent—*
   (*a*) *on the making of a bankruptcy order against him under Part IX of the Insolvency Act 1986 (1986 c. 45) or*
   (*b*) *on the sequestration of his estate under the Bankruptcy (Scotland) Act 1985 or when he grants a trust deed for his creditors.*

(5) *A company, partnership or individual shall also be treated as insolvent on the occurrence of any event corresponding to those specified in subsection (2), (3) or (4) under the law of Northern Ireland or of a country outside the United Kingdom.*

(6) *Where a provision is rendered ineffective by subsection (1), the parties are free to agree other terms for payment.*

(7) *In the absence of such agreement, the relevant provisions of the Scheme for Construction Contracts apply.*

How hard it is to define what is meant by 'is insolvent'?

When there is 'pay when paid' clause in the contract which is rendered ineffective by the Act it is likely that there will be no alternative payment provisions in the contract. In which case the whole of the payment regime of the Scheme applies.

The scheme for construction contracts [s114]

The Scheme is contained in a Regulation issued under the Act. The Act permits the Minister to make the Regulations after such consultation as he thinks fit. The Regulations themselves have to be approved by resolution of each House of Parliament.

114 The Scheme for Construction Contracts

(1) *The Minister shall by regulations make a scheme ('the Scheme for Construction Contracts') containing provision about the matters referred to in the preceding provisions of this Part.*

(2) *Before making any regulations under this section the Minister shall consult such persons as he thinks fit.*

(3) *In this section 'the Minister' means—*

(*a*) *for England and Wales, the Secretary of State, and*

(*b*) *for Scotland, the Lord Advocate.*

Where any provisions of the Scheme for Construction Contracts apply by virtue of this Part in default of contractual provision agreed by the parties, they have effect as implied terms of the contract concerned.

Regulations under this section shall not be made unless a draft of them has been approved by resolution of each House of Parliament.

Subsection 4 is very important. The Scheme applies as implied terms of the contract. Therefore both adjudication and the payment terms are contractual requirements and terms of the contract.

Service of notices [s115]

115 Service of notices, &c.

(1) *The parties are free to agree on the manner of service of any notice or other document required or authorised to be served in pursuance of the construction contract or for any of the purposes of this Part.*

(2) *If or to the extent that there is no such agreement the following provisions apply.*

(3) *A notice or other document may be served on a person by any effective means.*

(4) *If a notice or other document is addressed, pre-paid and delivered by post—*

(*a*) *to the addressee's last known principal residence or, if he is or has been carrying on a trade, profession or business, his last known principal business address, or*

(*b*) *where the addressee is a body corporate, to the body's registered or principal office,*

it shall be treated as effectively served.

(5) *This section does not apply to the service of documents for the purposes of legal proceedings, for which provision is made by rules of court.*

(6) *References in this Part to a notice or other document include any form of communication in writing and references to service shall be construed accordingly.*

Reckoning periods of time [s116]

The notice periods under the Act are very short. The appointment of the adjudicator and referral of the dispute to him is timetabled for 7 days. A day is a calendar day unless otherwise specified and calendar days include weekends. It is therefore essential that bank holidays be omitted from the calculations, but these are the only days omitted by the Act.

116 Reckoning of periods of time

(1) *For the purposes periods of time shall be reckoned as follows.*

(2) *Where an act is required to be done within a specified period after or from a specified date, the period begins immediately after that date.*

(3) *Where the period would include Christmas Day, Good Friday or a day which under the Banking and Financial Dealings Act 1971 (1971 c. 80) is a bank holiday in England and Wales or, as the case may be, in Scotland, that day shall be excluded.*

The construction industry has two recognised holiday periods — 2 weeks at Christmas and 1 week at Easter. These are not omitted from the time periods designated by the Act.

Crown application [s117]

117 Crown application

(1) *This Part applies to a construction contract entered into by or on behalf of the Crown otherwise than by or on behalf of Her Majesty in her private capacity.*

(2) *This Part applies to a construction contract entered into on behalf of the Duchy of Cornwall notwithstanding any Crown interest.*

(3) *Where a construction contract is entered into by or on behalf of Her Majesty in right of the Duchy of Lancaster. Her Majesty shall be represented, for the purposes of any adjudication or other proceedings arising out of the contract by virtue of this Part, by the Chancellor of the Duchy or such person as he may appoint.*

(4) *Where a construction contract is entered into on behalf of the Duchy of Cornwall, the Duke of Cornwall or the possessor for the time being of the Duchy shall be represented, for the Purposes of any adjudication or other proceedings arising out of the contract by virtue of this Part, by such person as he may appoint.*

Government contracts and the like are not exempt from the provisions of the Act.

That is the end of Part II of the Act.

14

The evolving legal position

General

The courts have generally been very supportive of adjudication recognizing that it was the will of Parliament that the process should produce a temporary decision enforceable by the courts. Adjudicators' decisions are enforceable summarily in the courts whether or not there has been procedural irregularity, or error.[1]

The general position is that provided the adjudicator acts within his jurisdiction and answers the questions that were put to him in the Notice of Adjudication then right or wrong his decision is enforceable.[2] In this the adjudicator's decision is similar to an expert determination.[3] Adjudication is a quasi-legal procedure[4] and therefore the decision must be based upon the legal rights of the parties. The adjudication clause, like an arbitration clause, is separable and even if the contract has been terminated disputes arising out of the contract can still be decided by an adjudicator.[5]

An adjudicator is not entitled to decide that he has too little information to make a determination.[6]

1 *Macob Civil Engineering* v. *Morrison Construction* (1999) 3BLR 93.

2 *Bouygues (UK) Limited* v. *Dahl-Jensen (UK) Limited* (2000) 1 BLR 49 & CA.

3 *Nikko Hotels (UK) Ltd* v. MEPC *Plc* (1991) 2 EGLR 103.

4 *A Straume (UK) Limited* v. *Bradlor Developments Limited* [1999] CILL 1520.

5 *A & D Maintenance Limited* v. *Pagehurst Construction Services Limited* (unreported).

6 *Ballast plc* v. *The Burrell Company (Construction Management) Limited* [21 June 2001].

The Court of Appeal has listed a number of salient features of adjudication:[7]

(*i*) A decision of an adjudicator whose validity is challenged as to its factual or legal conclusions or as to procedural error remains a decision that is both enforceable and should be enforced.

(*ii*) A decision that is erroneous, even if the error is disclosed by the reasons, will still not ordinarily be capable of being challenged and should, ordinarily, still be enforced.

(*iii*) A decision may be challenged on the ground that the adjudicator was not empowered by the Act to make the decision, because there was no underlying construction contract between the parties or because he had gone outside his terms of reference.

(*iv*) The adjudication is intended to be a speedy process in which mistakes will inevitably occur. Thus, the Court should guard against characterising a mistaken answer to an issue, which is within an adjudicator's jurisdiction, as being an excess of jurisdiction.

(*v*) It is a provisional decision, which, though enforceable at the time can be reopened on the final determination.

An issue as to whether a construction contract ever came into existence, which is one challenging the jurisdiction of the adjudicator, so long as it is reasonably and clearly raised, must be determined by the Court on the balance of probabilities with, if necessary, oral and documentary evidence.

The Notice of Adjudication

The Notice of Adjudication, which has to be in writing,[8] sets out the jurisdiction of the adjudicator.[9] There must be a dispute between the parties for the adjudicator to have jurisdiction and a dispute means whatever claims, heads of claim, issues, contentions

7 *C & B Scene Concept Design Ltd* v. *Isobars Ltd*. GWS.

8 *Strathmore Building Services Limited* v. *Colin Scott Grig t/a Hestia Fireside Services* [SWS].

9 *Fastrack Contractors Limited* v. *Morrison Construction Ltd and Impregilo UK Ltd* [2000] BLR 168.

or causes of action that were in dispute at the moment that the referring party first intimated an adjudication reference and chose to crystallize into an adjudication reference.[10] However, a Scottish court hinted that this definition might be too wide[11] and for the Scheme the issues have to be related to the generality of the dispute referred.[12] The dispute consists of the arguments between the parties at the time of referral and not further arguments that have not already been rehearsed between the parties.[13]

The Notice of Adjudication must comply with the contract requirements and describe precisely the disputes being referred and only those so described will be enforced.[14] However, when the Notice of Adjudication is not precise about the disputes being referred a broad, rather than a narrow, construction should be put on it,[15] the court taking the view that these documents are drafted by practitioner rather than lawyers, but the same latitude ought not to be given to one drafted with the benefit of outside advice.[16]

A Notice of Adjudication may be given before a payment obligation arises. If the paying party does not pay there is a dispute. Just as a court can give a declaration as to the future, an adjudicator can do also.[17]

Even if a dispute is the subject of existing proceedings in court it can still be referred to an adjudicator and enforced pending a decision by the court. There is no provision in the HGC&R Act for a stay of execution.[18] A party does not waive its right to

---

10 *Fastrack Contractors Limited* v. *Morrison Construction Ltd and Impregilo UK Ltd* [2000] BLR 168.

11 *Barr Limited* v. *Law Mining Limited* [15 June 2001 SWS].

12 *Chamberlain Carpentry & Joinery Ltd* v. *Alfred McAlpine Construction Limited* (21 March 2002 BAILII).

13 *Edmund Nuttall Limited* v. *R.G. Carter Limited* (21 March 2002 BAILII).

14 *Ken Griffin & John Tomlinson t/a K&D Contractors* v. *Midas Homes Limited* [unreported].

15 *Karl Construction (Scotland) Limited* v. *Sweeney Civil Engineering (Scotland) Ltd* [21 December, 2000 SWS].

16 *Ken Griffen* v. *Midas Homes Ltd* (2000 unreported).

17 *Discain Project Services* v. *Opecprime Developments Limited* (2000) 8 BLR 402.

18 *A & D Maintenance Limited* v. *Pagehurst Construction Services Limited* (unreported).

adjudication because it has started court proceedings and a stay of the adjudication will be refused. However, if there is evidence at the enforcement proceedings that the payee is in liquidation,[19] or in receivership[20] or might have been unable to repay the money if the appeal went against him[21] a stay may be granted but might be lifted if security is provided. Similarly, if there is an arbitration clause in the contract a stay will not be granted pending the outcome of the arbitration.[22]

When a party is ordered by an adjudicator to make a payment fails to do so and the payee starts proceedings for enforcement the paying party will be liable for the payee's costs in making the application even if the sum is paid in full prior to the court hearing.[23]

A project manager can enter into a construction contract as agent for an employer, and then the provisions of that Act apply to that contract between the contractor and the employer.[24]

The terms of the Housing Grants, Construction & Regeneration Act Part II are mandatory and cannot be contracted out of.[25]

Adjudication does not contravene s6 of the Human Rights Act.[26,27]

The most fruitful challenges have been mounted against enforcement on the grounds of lack of jurisdiction.

Jurisdiction

As with all forms of dispute resolution it is often in one party's perceived interest that the adjudication should not proceed.

19 *Bouygues (UK) Limited* v. *Dahl-Jensen (UK) Limited* (2000) 1 BLR 49 & CA.

20 *Rainford House Limited* v. *Cadogan Limited* (2001 unreported).

21 *Hertschel Engineering Limited* v. *Breen Property Limited.* [2000 GWS].

22 *Absolute Rentals Limited* v. *Gencor Enterprises Limited* (a company registered in Cyprus) [2000 CILL 1637].

23 *Outwing Construction Limited* v. *H Randell & Son Limited* [1999] BLR 156.

24 *Universal Music Operations Limited* v. *Flairnote Limited & others* [unreported].

25 *Christiani & Nielsen Limited* v. *The Lowry Centre Development Company Limited* (unreported).

26 *Elanay Contract Ltd* v. *The Vestry* [2000 CILL 1679].

27 *Austin Hall Building Limited* v. *Buckland Securities Limited* [Apr. 2001 GWS].

One method of attempting to delay the adjudication is to apply for an injunction to restrain a party from initiating an allegedly void reference and pursing it but there does not appear to be any legal or equitable interest at that stage that an injunction would protect.[28] When an adjudication has started and a court holds, before a decision is reached, that the adjudicator lacks jurisdiction an injunction may be granted to restrain the referring party from taking any substantial step in adjudication or seeking to enforce or implement any award which may be made.[29] When it is common ground that a contract exists between the parties, which incorporates DOM/1, to which the Act applies, the court does not need to investigate any further and an injunction would be refused.[30]

It is clear that an adjudicator has no power to decide his own jurisdiction[31] although the parties can agree to give him that power.[32] Nevertheless if his jurisdiction is challenged he has to consider it. He appears to have three options: he can ignore the challenge and proceed and leave it to a party to challenge his jurisdiction at enforcement proceedings; he can investigate the issue and decide that he has jurisdiction in which case he would proceed; or he can decide he had no jurisdiction and decline to continue.[33] Where a decision is subsequently challenged and the lack of jurisdiction is clearly arguable then the court may decline to enforce the decision pending its judgement on the jurisdiction issue.[34]

All the terms, and not merely the existence, of a construction contract have to be evidenced in writing if the contract is to be capable of being referred to adjudication under the Housing Grants, Construction and Regeneration Act 1996.[35]

---

28 *Workplace Technologies Plc* v. *E Squared Technologies Limited and Another* [2000 CILL 1607].

29 *John Mowlem & Company plc* v. *Hydra-Tight Ltd.*

30 *R.G. Carter Limited* v. *Edmund Nuttall Limited* (21 June 2000).

31 *The Project Consultancy Group* v. *The Trustees of the Gray Trust* [1999] BLR 377.

32 *Nolan Davis Limited* v. *Steven P Catton* (unreported).

33 *Christiani & Nielsen* v. *The Lowry Development Company Ltd* (unreported).

34 *The Project Consultancy Group* v. *The Trustees of the Gray Trust* [1999] BLR 377.

35 *RJT Consulting Engineers Ltd* v. *DM Engineering (Northern Ireland)* Ltd (2002).

When the issue is whether or not a contract contains a contract period (i.e. greater or less than 45 days) which permits interim payment under s109 it is a matter that the adjudicator can decide.[36]

Settlement agreements are outwith the jurisdiction of an adjudicator because they are not construction contracts.[37] So are matters which precede the making of the contract — such questions as to whether the contract was entered into on a false basis, such as economic distress or as a result of a misrepresentation or on the basis of promises which did not materialise because none of them were disputes under the contract.[38]

Any points which are to be made as to whether the Adjudicator has jurisdiction must be raised in a timely manner during the adjudication itself, even if no contract existed and the referral was not under the Act, because then the adjudication took place by agreement between the parties on the same terms as the Act and the Scheme.[39]

Procedural matters

Although an adjudicator has the right to conduct an entirely inquisitorial process it does not follow that he is bound to consider alternatives when submissions have been made to him.[40]

Natural justice

The *Macob* position has now been almost completely eroded. It is accepted that the adjudicator has to conduct the proceedings in accordance with the rules of natural justice or as fairly as the limitations imposed by Parliament permit.[41] There is no reason in law why the adjudicator should not have individual telephone

36 *Tim Butler Contractors Limited* v. *Merewood Homes Limited* [unreported].

37 *Lathom Construction Limited* v. *Brian Cross and Ann Cross* CILL 1565.

38 *Shepherd Construction Limited* v. *Mecright Ltd* [2000] BLR 489.

39 *Maymac Environmental Services Ltd* v. *Faraday Building Services Ltd* [2000 CILL 1686].

40 *Allied London and Scottish Properties plc* v. *Riverbrae Limited* [1999] BLR 346.

41 *Glencot Devopment and Design* v. *Ben Barrett & Son (Contractors) Ltd* [unreported].

conversations with the parties.[42] Whilst the adjudicator may, due to time restraints, speak to the parties separately he should advise the other party himself of what he has learnt that may be relevant to his decision and not leave it to the party to inform that other party of what he told the adjudicator.[43] The adjudicator may not take evidence from third parties and not give the parties an opportunity to comment on it,[44] nor should he give a witness statement for use in enforcement proceedings[45] because he would be seen not to be acting impartially.

However, it seems that the adjudicator may ascertain points of law and decide them without reference to the parties, even if the point has not been raised by either of them.[46] There seems to be a difference between investigating fact and law. Acting inquisitorially may mean making telephone calls but does not mean acting unfairly. There is a difference between telephone calls of a purely administrative nature and calls that convey or elicit 'relevant information'. The former would be sensible for the adjudicator to have made by a secretary; the latter should be made with great caution.[47]

An adjudicator should not try to act as a mediator even if the parties agree to that procedure. Such agreement does not deprive a party of the right to claim apparent bias of the adjudicator, which would render the decision unenforceable.[48]

The fundamental point is that the adjudicator must be seen to act impartially and fairly, but it is looking increasingly likely that courts will implement a supervisory role on the conduct of adjudications.

---

42 *Discain Project Services* v. *Opecprime Developments* [2] [26/4/2001 GWS].
43 *Discain Project Services* v. *Opecprime Developments* (2000) 8 BLR 402.
44 *Woods Hardwick Ltd* v. *Chiltern Air Conditioning* (7 July 2000 GWS).
45 *Discain Project Services* v. *Opecprime Developments Limited* (2000) 8 BLR 402.
46 *Karl Construction (Scotland) Limited* v. *Sweeney Civil Engineering (Scotland) Ltd* [21 December, 2000 SWS].
47 *Discain Project Services* v. *Opecprime Developments* [2] [26/4/2001 GWS].
48 *Glencot* v. *Barrett* (2001 unreported).

Adjudicators' fees and costs of the adjudication

An adjudicator's fees fall within the words 'anything done or omitted in the discharge or purported discharge of his functions as adjudicator.' These fees may be challenged if, and only if, the adjudicator has acted in bad faith. The criterion of the reasonably competent solicitor is not a relevant test in determining the reasonableness of an adjudicator's fees. A court must be very slow indeed to substitute its own view of what constitutes reasonable hours.[49]

If the adjudicator is an employee of a company it is acceptable for the company to invoice for and collect his fees.[50]

Misdescription of one of the parties to an adjudication did not deprive the adjudicator of jurisdiction where the parties had not been misled. The costs of an adjudication could not be recovered as damages. It was not the function of the court in enforcement proceedings to review each and every aspect of the adjudicator's award.[51]

The Scheme

References under the scheme are limited to disputes arising out of the contract. Matters of jurisdiction, and misrepresentation do not arise under the contract.

A list of prospective adjudicators specified in a contract from which the selection is to be made is acceptable.[52]

Where a Notice of Adjudication, which referred to various earlier invoices and letters, failed to describe precisely which disputes had been referred only the disputes defined in accordance with the Scheme came within the adjudicator's jurisdiction.[53]

---

49 *Stubbs Rich Architects* v. *WH Tolley & Son Limited* (8 August 2001).

50 *Faithful & Gould Limited* v. *Arcal Limited and Ors* (25 May 2001).

51 *Total M&E Services Ltd* v. *ABB Building Technologies Ltd (formally ABB Steward Ltd)*.

52 *John Mowlem & Company plc* v. *Hydra-Tight Ltd*.

53 *Ken Griffin & John Tomlinson t/a K&D Contractors* v. *Midas Homes Limited* [unreported].

Paragraph 8 indicates that it is only with the consent of the parties that an adjudicator can adjudicate at the same time on more than one dispute or adjudicate under related disputes under different contracts,[54] which effectively means that appointing authorities may not appoint the same adjudicator for disputes suitable for 'joinder'. However, one dispute can be all or part of whatever is in dispute at the moment that the referring party first intimates an adjudication reference,[55] and this may consist of many issues. What comprises a single dispute is a matter of fact to be addressed objectively for each case[56] and a dispute as to the costs of the adjudication is not another dispute but an aspect of the first.

The award of parties' costs is not covered by the Scheme but only the determination and apportionment of the adjudicator's own costs and expenses.[57] It was held in an early case that there was an implied term in the Scheme which permitted the adjudicator to allocate the parties' costs[58] but this was subsequently contradicted when it was held that there was an implied agreement between the parties authorising the adjudicator to allocate parties' costs when both parties applied in their submissions for their costs to be paid.[59] This latter position has been followed.[60]

For the purposes of paragraph 9 when there is an allegation that two disputes are substantially the same the court may embark on a limited enquiry to see whether there is substantial overlap between the two respective adjudicator's adjudications and decisions, but not into the merits of the disputes.[61]

---

54 *Grovedeck Limited* v. *Capital Demolitions Limited* [2000] BLR 181.

55 *Fastrack Contractors Limited* v. *Morrison Construction Ltd and Impregilo UK Ltd* [2000] BLR 168.

56 *Chamberlain Carpentry & Joinery Ltd* v. *Alfred McAlpine Construction Ltd* (25 March 2002 BAILLI).

57 Paragraph 25 of the Scheme for Construction Contracts (England and Wales) Regulations 1998 No. 649.

58 *John Cothliff Limited* v. *Allen Build (North West) Limited* [199] BLR 426.

59 *Northern Developments (Cumbria) Ltd* v. *J&J Nichol* [2000] CILL 1601 and GWS.

60 *Nolan Davis Limited* v. *Steven P Catton* (unreported).

61 *Sherwood & Casson Limited* v. *Mackenzie Engineering Limited* [2000] CILL 1577.

Mistakes

If an adjudicator makes a mistake in his decision, whether in fact or law,[62] it will be enforced provided that he has not exceeded his jurisdiction,[63] that is because if he exceeds his jurisdiction he has not been authorized to make a mistake.[64]

In the absence of an express term it will be an implied term that the adjudicator can correct clerical or arithmetical errors in his decision provided that he does so within a reasonable time and without prejudice to the other party.[65,66]

Parties' costs

The recovery of parties' costs is a matter for the contract and the Scheme is silent on the subject (see also under Scheme). The parties appear free to reach their own decision on what costs may be recovered and a provision amending the CIC Model Adjudication Procedure to the effect that the referring party pays the costs of both parties and the adjudicator in any event was held to be valid.[67] However, a successful party cannot claim its costs of adjudication as damages for breach of contract and unless the adjudicator is given jurisdiction to award parties' costs and does so such costs cannot be recovered.[68]

Payment provisions

When a payment is dependent upon a certificate, a procedure which does not give a time period within which certification is to be made does not comply with s110(1)(a) of the Act.[69]

---

62 *SL Timber Systems Limited* v. *Carillion Construction Limited* [2000] (27 June).

63 *Bouygues (UK) Limited* v. *Dahl-Jensen (UK) Limited* (2000) 1 BLR 49 & CA.

64 *FW Cook Ltd* v. *Shimzu (UK) Ltd.*

65 *Bloor Construction (UK) Ltd* v. *Bowmer & Kirkland (London) Limited* [2000] BLR 764.

66 *Edmund Nuttall Ltd* v. *Sevenoaks District Council* [unreported].

67 *Bridgeway Construction* v. *Tolent Construction Limited* [unreported].

68 *Total M&E Services* v. *ABB Technologies Ltd* (26 Feb. 02 HHJ Wilcox unreported) "the provisional nature of adjudication, which, though enforceable at the time can be reopened on the final determination".

69 *Maxi Construction Management Limited* v. *Morton Rolls Limited* [SWS 7 August 2001].

A notice under s111 had to be in writing.[70] A notice of intention to withhold payment under s111 to be arguably effective has to precede the reference to adjudication and provide a comprehensive code governing the right of set off against payments contractually due.[71]

A withholding notice issued under s111(1) is ineffective and invalid unless it specifies the amount it is proposed to withhold,[72] however it is possibly that a notice referring back to an earlier letter might suffice.[73] A notice under s110 does not establish a right but bars an existing right if not given. A notice under s110 was required before monies could be withheld for either set off or abatement.[74] No notice is required where the work has been overmeasured because the sum claimed was never due.

Failure to give either a 110(2) or a 111 notice does not in any way preclude a dispute about an amount claimed and it is incumbent upon the claimant to demonstrate that the sum claimed is contractually due. However, withholding an overpayment on another contract is a set off and requires a withholding notice.[75]

The terms set off and abatement are confusing. Abatement 'applies only to matters that go to reduce the value of the work performed or the goods sold, it cannot be applied to a cross-claim for delay in the execution of the work, which would be a matter of equitable set off'.[76] 'An arguable cross-claim whose amount is known or can be ascertained with certainty can be set off as a defence even if the defendant's right to payment of the liquidated amount is disputed'.[77] Abatement therefore goes to the heart of

70 *Strathmore Building Services Limited* v. *Colin Scott Grieg t/a Hestia Fire Design* [SWS].

71 *VHE Construction plc* v. *RBSTB Trust Co Limited* [2000] BLR 187.

72 *Palmers Limited* v. *ABB Power Construction Limited* [1999] BLR 426.

73 *Strathmore Building Services Limited* v. *Colin Scott Grieg t/a Hestia Fire Design* [SWS].

74 *Whiteways Contractors (Sussex) Limited* v. *Impresa Castelli Construction UK Ltd* WS TCC.

75 *SLD Timber Systems* v. *Carillion Construction* [SWS 30 July 2001].

76 *Keating on Building Contracts* 7th Edition 17–66 ex *Mellowes Archital Ltd* v. *Bell Projects Ltd* 87 BLR 26, CA.

77 *Keating on Building Contracts* 7th Edition 17–65 ex *Axel Johnson* v. *Mineral Group* [1992 1 WLR270, CA].

what is meant by the sum due under the contract. Some reductions may be taken into account by the adjudicator without a withholding notice in such matters as overmeasurement or overstating the value of the work in an application and abatement for defective work notified at the time of the application because these were sums that were never due.

It is not an implied term of the contract between the parties that liquidated and ascertained damages could be deducted from a payment following a decision of an adjudicator where a claim to deduct those damages had not been made at the relevant time in accordance with the contractual machinery.[78]

Interpretation of the Act

The statute is not to be construed by reference to the statutory instrument made under it.[79] A novation agreement for a construction contract falls within s104 even if the original contract was formed prior to 1 May 1998.[80]

s104(2) includes an agreement to do architectural, design or surveying work in relation to construction operations. Works of a surveyor producing evidence as an expert in arbitration proceedings is not work in relation to construction operations.[81]

s105(2) which defines the exemptions from construction operations, has caused uncertainty. Scaffolding was held to be a construction operation although the work it was provided for (the erection of a boiler) was not.[82] A contract for the annual service and maintenance and comprising a responsive breakdown service for heating systems, gas fires and cookers in the residential properties owned by a local authority was held to be a construction contract

78 *Edmund Nuttall Ltd* v. *Sevenoaks District Council* (14 April 2000).

79 *Grovedeck Limited* v. *Capital Demolitions Limited* [2000] BLR 181.

80 *Yarm Road Limited* v. *Costain Ltd* [30 July 2001].

81 *Fence Gate Limited* v. *James R Knowles Limited (Technology and Construction Court, Salford)* HHJ Gilliland QC — 31 May 2001.

82 *Palmers Limited* v. *ABB Power Construction Limited* [1999] BLR 426.

within s105(1)(a & c).[83] For determining the 'primary activity' on the site for the purposes of s105(2) it may not be confined to those activities taking place within the definition of the 'site' as defined in the contract.[84] Thus for a contract for the installation of a generator for printing works the primary activity was held to be printing and not power generation. Where the contract work falls within s105(2)(c) (i)&(ii), pipework (electrical connections, conduits, etc.) which connects items of plant it is exempt because it is 'in a real sense part of the plant being assembled or installed'.[85] Steam boilers installed on a separate site but generating steam for an oil refinery also fell within s105(2)(c).[86]

The words to 'form part of the land' in s104 have been subject to scrutiny. The assembling of a boiler on site and then erecting into position was held to be a construction operation,[87] but the assembly of a module in a yard at Teeside for accommodation on an oil rig in the Gulf of Mexico was held neither to form part of the land at Teeside nor in the Gulf of Mexico because the land had to be in England, Wales or Scotland.[88] A situation below low water line would also not form part of the land. Neither do some shop fittings.[89]

A mediation precondition undermines the right of either party to refer a dispute to adjudication 'at any time' (s108) and therefore is not compliant with the Act.[90]

An oral contract becomes a contract in writing in accordance with s107(2)(c) when it is evidenced in writing. However, to be evidenced in writing there must be evidence of all the terms being in writing, or at least the terms relevant to the matter is dispute.[91]

83 *Nottingham Community Housing Association* v. *Powerminster Limited* [GWS].
84 *ABB Zantingh Limited* v. *Zedal Building Services Ltd* [GWS 1/12/00].
85 *Homer Burgess Ltd* v. *Chired (Annan) Ltd* [2000] BLR 124.
86 *Mitsui Babcock Energy Services Limited* [SWS 13 June 2001].
87 *Palmers Limited* v. *ABB Power Construction Limited* [1999] BLR 426.
88 *Staveley Industries plc* v. *Odebrecht Oil & Gas Services Ltd* (unreported).
89 *Gibson Lea Retail Interiors Limited* v. *Makro Self-Services Wholesalers* [24 July, 2001].
90 *R. G. Carter Limited* v. *Edmund Nuttall Limited* (21 June 2000 BAILII).
91 *RJT Consulting Engineers Ltd* v. *DM Engineering (N.I.) Ltd* (CA) [GWS].

Under s107(5) if the existence of an oral contract is not denied 'in adjudication proceedings, or in arbitral or legal proceedings' it is evidenced in writing. This was held to refer to being denied in 'other preceding adjudication proceedings'.[92] This decision revolved about the meaning of the words 'to the effect alleged', which means the terms of the oral contract alleged.

Where approximately 65% of a contract is for work which relates to operations on a dwelling which one of the parties to the contract occupies or intends to occupy as his residence the contract is not excluded from the Act by s106.[93]

Failure to serve any notice under s110(2) or a timeous notice under s111 of the Housing Grants, Construction and Regeneration Act 1996 did not debar a defendant from disputing that the amount claimed by the claimant was not a 'sum due' under the relevant contract. However, an adjudicator's decision to the contrary did not mean that his award was outside his jurisdiction. An averment that the claimant was insolvent did not constitute a defence.[94]

Bankruptcy

Adjudication is a legal procedure for the purposes of s11(3) of the Insolvency Act 1983. A stay might be granted if court proceedings had commenced and there was evidence that the receiving party might not be able to repay the sum if the court decision went against him,[95] or that payment would put the paying party into liquidation.[96,97]

Enforcement

From the first reported case[98] the courts have enforced adjudicators' decisions summarily. The Court of Appeal in all its adjudication judgements have expressed the intention that adjudicator's decisions shall normally be enforced. 'I would regard it as a pity if too

92 *Grovedeck Limited* v. *Capital Demolitions Limited* [2000] BLR 181.
93 *Samuel Thomas Construction Limted* v. *Bick and Bick* [unreported].
94 *SLTimber Systems Ltd* v. *Carillion Construction Ltd.*
95 *Hertschel Engineering Limited* v. *Breen Property Limited* [2000 GWS].
96 *Holt* v. *Colt* (unreported).
97 *George Parke* v. *Fenton Gretton Partnership* [2 August 2000].
98 *Macob Civil Engineering* v. *Morrison Construction* (1999) 3BLR 93.

much "jurisdictional wrangling" were to limit the opportunities for expeditious adjudication'.[99]

The ICE contracts and NEC

The notice of dissatisfaction procedure was stated to be illegal but the case had not been argued.[100] It is clearly wrong. It may not be Act-compliant but certainly not illegal. The question is whether you can define in a contract when a dispute arises. The argument against is that there is a definition in law concerning what constitutes a dispute,[101] and argument in favour is that if Parliament allows in the Scheme for the contracting parties to determine when disputes cannot arise (i.e. when the decision is said to be final and conclusive) and there does not appear, therefore, to be any reason why the parties should not be able to do likewise in their adjudication provision.

Glossary

GWS Government Court Services Web site
www.courtservice.gov.uk/judgments/judg_home.htm

SWS Scottish Web Site
www.scotcourts.gov.uk

CILL Construction Industry Law Letter published by LLP and edited by Tony Francis, Fenwick Elliott

BAILII British and Irish Legal Institute
www.bailii.org/

[unreported] These cases are not reported in any standard publication but they may be found on one of the following websites:
www.winwardfearon.co.uk/adjudication.htm
www.adjudication.co.uk/

99 Lord Justice Ward in *RJT Consulting Engineers Ltd* v. *DM Engineering (N.I.) Ltd* (CA) (www.adjudication.co.uk).

100 *John Mowlem & Company plc* v. *Hydra-Tight Ltd.*

101 *Halki Shipping* v. *Sopex Oils* [1998 1 WLR 726].

Appendix A

Part II: Housing Grants, Construction and Regeneration Act 1996

The text of Part II of the Housing Grants, Construction and Regeneration Act 1996 is reproduced below.

Part II Construction contracts

Introductory provisions

104 Construction contracts

(1) In this Part a "construction contract" means an agreement with a person for any of the following—

(*a*) the carrying out of construction operations;

(*b*) arranging for the carrying out of construction operations by others, whether under subcontract to him or otherwise;

(*c*) providing his own labour, or the labour of others, for the carrying out of construction operations.

(2) References in this Part to a construction contract include an agreement—

(*a*) to do architectural, design or surveying work, or

(*b*) to provide advice on building, engineering, interior or exterior decoration or on the laying-out of landscape,

in relation to construction operations.

(3) References in this Part to a construction contract do not include a contract of employment (within the meaning of the Employment Rights Act 1996 (1996 c.18)).

(4) The Secretary of State may by order add to, amend or repeal any of the provisions of subsection (1), (2) or (3) as to the

agreements which are construction contracts for the purposes of this Part or are to be taken or not to be taken as included in references to such contracts.

No such order shall be made unless a draft of it has been laid before and approved by a resolution of each of (*sic*) House of Parliament.

(5) Where an agreement relates to construction operations and other matters, this Part applies to it only so far as it relates to construction operations.

An agreement relates to construction operations so far as it makes provision of any kind within subsection (1) or (2).

(6) This Part applies only to construction contracts which—

(*a*) are entered into after the commencement of this Part, and

(*b*) relate to the carrying out of construction operations in England, Wales or Scotland.

(7) This Part applies whether or not the law of England and Wales or Scotland is otherwise the applicable law in relation to the contract.

105 Meaning of "construction operations"

(1) In this Part "construction operations" means, subject as follows, operations of any of the following descriptions—

(*a*) construction, alteration, repair, maintenance, extension, demolition or dismantling of buildings, or structures forming or to form, part of the land (whether permanent or not);

(*b*) construction, alteration, repair, maintenance, extension, demolition or dismantling of any works forming, or to form, part of the land, including (without prejudice to the foregoing) walls, roadworks, power-lines, telecomunication apparatus, aircraft runways, docks and harbours, railways, inland waterways, pipe-lines, reservoirs, water-mains, wells, sewers, industrial plant and installations for purposes of land drainage, coast protection or defence;

(*c*) installation in any building or structure of fittings forming part of the land, including (without prejudice to the foregoing) systems of heating, lighting, air-conditioning, ventilation, power supply, drainage, sanitation, water supply or fire protection, or security or communications systems;

(*d*) external or internal cleaning of buildings and structures, so far as carried out in the course of their construction, alteration, repair, extension or restoration;

(*e*) operations which form an integral part of, or are preparatory to, or are for rendering complete, such operations as are previously described in this subsection, including site clearance, earth-moving, excavation, tunnelling and boring, laying of foundations, erection, maintenance or dismantling of scaffolding, site restoration, landscaping and the provision of roadways and other access works;

(*f*) painting or decorating the internal or external surfaces of any building or structure.

(2) The following operations are not construction operations within the meaning of this Part—

(*a*) drilling for, or extraction of, oil or natural gas;

(*b*) extraction (whether by underground or surface working) of minerals; tunnelling or boring, or construction of underground works, for this purpose;

(*c*) assembly, installation or demolition of plant or machinery, or erection or demolition of steelwork for the purposes of supporting or providing access to plant or machinery, on a site where the primary activity is—

(*i*) nuclear processing, power generation, or water or effluent treatment, or

(*ii*) the production, transmission, processing or bulk storage (other than warehousing) of chemicals, pharmaceuticals, oil, gas, steel or food and drink;

(*d*) manufacture or delivery to site of—

(*i*) building or engineering components or equipment,

(*ii*) materials, plant or machinery, or

(*iii*) components for systems of heating, lighting, air-conditioning, ventilation, power supply, drainage, sanitation, water supply or fire protection, or for security or communications systems, except under a contract which also provides for their installation;

(*e*) the making, installation and repair of artistic works, being sculptures, murals and other works which are wholly artistic in nature.

(3) The Secretary of State may by order add to, amend or repeal any of the provisions of subsection (1) or (2) as to the operations and work to be treated as construction operations for the purposes of this Part.

(4) No such order shall be made unless a draft of it has been laid before and approved by a resolution of each House of Parliament.

106 Provisions not applicable to contract with residential occupier

(1) This Part does not apply—

(*a*) to a construction contract with a residential occupier (see below), or

(*b*) to any other description of construction contract excluded from the operation of this Part by order of the Secretary of State.

(2) A construction contract with a residential occupier means a construction contract which principally relates to operations on a dwelling which one of the parties to the contract occupies, or intends to occupy, as his residence.

In this subsection "dwelling" means a dwelling-house or a flat; and for this purpose—

"dwelling-house" does not include a building containing a flat; and

"flat" means separate and self-contained premises constructed or adapted for use for residential purposes and forming part of a building from some other part of which the premises are divided horizontally.

(3) The Secretary of State may by order amend subsection (2).

(4) No order under this section shall be made unless a draft of it has been laid before and approved by a resolution of each House of Parliament.

107 Provisions applicable only to agreements in writing

(1) The provisions of this Part apply only where the construction contract is in writing, and any other agreement between the parties as to any matter is effective for the purposes of this Part only if in writing.

The expressions "agreement", "agree" and "agreed" shall be construed accordingly.

(2) There is an agreement in writing—

(*a*) if the agreement is made in writing (whether or not it is signed by the parties),

(*b*) if the agreement is made by exchange of communications in writing, or

(*c*) if the agreement is evidenced in writing.

(3) Where parties agree otherwise than in writing by reference to terms which are in writing, they make an agreement in writing.

(4) An agreement is evidenced in writing if an agreement made otherwise than in writing is recorded by one of the parties, or by a third party, with the authority of the parties to the agreement.

(5) An exchange of written submissions in adjudication proceedings, or in arbitral or legal proceedings in which the existence of an agreement otherwise than in writing is alleged by one party against another party and not denied by the other party in his response constitutes as between those parties an agreement in writing to the effect alleged.

(6) References in this Part to anything being written or in writing include its being recorded by any means.

Adjudication

108 Right to refer disputes to adjudication

(1) A party to a construction contract has the right to refer a dispute arising under the contract for adjudication under a procedure complying with this section.

For this purpose "dispute" includes any difference.

(2) The contract shall—

(*a*) enable a party to give notice at any time of his intention to refer a dispute to adjudication;

(*b*) provide a timetable with the object of securing the appointment of the adjudicator and referral of the dispute to him within seven days of such notice;

(*c*) require the adjudicator to reach a decision within 28 days of referral or such longer period as is agreed by the parties after the dispute has been referred;

(*d*) allow the adjudicator to extend the period of 28 days by up to 14 days, with the consent of the party by whom the dispute was referred;

(*e*) impose a duty on the adjudicator to act impartially; and

(*f*) enable the adjudicator to take the initiative in ascertaining the facts and the law.

(3) The contract shall provide that the decision of the adjudicator is binding until the dispute is finally determined by legal proceedings, by arbitration (if the contract provides for arbitration or the parties otherwise agree to arbitration) or by agreement.

The parties may agree to accept the decision of the adjudicator as finally determining the dispute.

(4) The contract shall also provide that the adjudicator is not liable for anything done or omitted in the discharge or purported discharge of his functions as adjudicator unless the act or omission is in bad faith, and that any employee or agent of the adjudicator is similarly protected from liability.

(5) If the contract does not comply with the requirements of subsections (1) to (4), the adjudication provisions of the Scheme for Construction Contracts apply.

(6) For England and Wales, the Scheme may apply the provisions of the Arbitration Act 1996 (1996 c. 23) with such adaptations and modifications as appear to the Minister making the scheme to be appropriate.

For Scotland, the Scheme may include provision conferring powers on courts in relation to adjudication and provision relating to the enforcement of the adjudicator's decision.

Payment

109 Entitlement to stage payments

(1) A party to a construction contract is entitled to payment by instalments, stage payments or other periodic payments for any work under the contract unless—
   (*a*) it is specified in the contract that the duration of the work is to be less than 45 days, or
   (*b*) it is agreed between the parties that the duration of the work is estimated to be less than 45 days.
(2) The parties are free to agree the amounts of the payments and the intervals at which, or circumstances in which, they become due.
(3) In the absence of such agreement, the relevant provisions of the Scheme for Construction Contracts apply.
(4) References in the following sections to a payment under the contract include a payment by virtue of this section.

110 Dates for payment

(1) Every construction contract shall—
   (*a*) provide an adequate mechanism for determining what payments become due under the contract, and when, and
   (*b*) provide for a final date for payment in relation to any sum which becomes due.

   The parties are free to agree how long the period is to be between the date on which a sum becomes due and the final date for payment.
(2) Every construction contract shall provide for the giving of notice by a party not later than five days after the date on which a payment becomes due from him under the contract, or would have become due if—
   (*a*) the other party had carried out his obligations under the contract, and

(*b*) no set-off or abatement was permitted by reference to any sum claimed to be due under one or more other contracts,

specifying the amount (if any) of the payment made or proposed to be made, and the basis on which that amount was calculated.

(3) If or to the extent that a contract does not contain such provision as is mentioned in subsection (1) or (2), the relevant provisions of the Scheme for Construction Contracts apply.

111 Notice of intention to withhold payment

(1) A party to a construction contract may not withhold payment after the final date for payment of a sum due under the contract unless he has given an effective notice of intention to withhold payment.

The notice mentioned in section 110(2) may suffice as a notice of intention to withhold payment if it complies with the requirements of this section.

(2) To be effective such a notice must specify—

(*a*) the amount proposed to be withheld and the ground for withholding payment, or

(*b*) if there is more than one ground, each ground and the amount attributable to it,

and must be given not later than the prescribed period before the final date for payment.

(3) The parties are free to agree what that prescribed period is to be.

In the absence of such agreement, the period shall be that provided by the Scheme for Construction Contracts.

(4) Where an effective notice of intention to withhold payment is given, but on the matter being referred to adjudication it is decided that the whole or part of the amount should be paid, the decision shall be construed as requiring payment not later than—

(*a*) seven days from the date of the decision, or

(*b*) the date which apart from the notice would have been the final date for payment,

whichever is the later.

112 Right to suspend performance for non-payment

(1) Where a sum due under a construction contract is not paid in full by the final date for payment and no effective notice to withhold payment has been given, the person to whom the sum is due has the right (without prejudice to any other right or remedy) to suspend performance of his obligations under the contract to the party by whom payment ought to have been made ("the party in default").

(2) The right may not be exercised without first giving to the party in default at least seven days' notice of intention to suspend performance, stating the ground or grounds on which it is intended to suspend performance.

(3) The right to suspend performance ceases when the party in default makes payment in full of the amount due.

(4) Any period during which performance is suspended in pursuance of the right conferred by this section shall be disregarded in computing for the purposes of any contractual time limit the time taken, by the party exercising the right or by a third party, to complete any work directly or indirectly affected by the exercise of the right.

Where the contractual time limit is set by reference to a date rather than a period, the date shall be adjusted accordingly.

113 Prohibition of conditional payment provisions

(1) A provision making payment under a construction contract conditional on the payer receiving payment from a third person is ineffective, unless that third person, or any other person payment by whom is under the contract (directly or indirectly) a condition of payment by that third person, is insolvent.

(2) For the purposes of this section a company becomes insolvent—

(*a*) on the making of an administration order against it under Part II of the Insolvency Act 1986 (1986 c. 45),

(*b*) on the appointment of an administrative receiver or a receiver or manager of its property under Chapter I of Part III of that Act, or the appointment of a receiver under Chapter II of that Part, (*c*) on the passing of a resolution for voluntary winding-up without a declaration of solvency under section 89 of that Act, or

(*d*) on the making of a winding-up order under Part IV or V of that Act.

(3) For the purposes of this section a partnership becomes insolvent—

(*a*) on the making of a winding-up order against it under any provision of the Insolvency Act 1986 as applied by an order under section 420 of that Act, or

(*b*) when sequestration is awarded on the estate of the partnership under section 12 of the Bankruptcy (Scotland) Act 1985 (1985 c. 66) or the partnership grants a trust deed for its creditors.

(4) For the purposes of this section an individual becomes insolvent—

(*a*) on the making of a bankruptcy order against him under Part IX of the Insolvency Act 1986 (1986 c. 45), or

(*b*) on the sequestration of his estate under the Bankruptcy (Scotland) Act 1985 or when he grants a trust deed for his creditors.

(5) A company, partnership or individual shall also be treated as insolvent on the occurrence of any event corresponding to those specified in subsection (2), (3) or (4) under the law of Northern Ireland or of a country outside the United Kingdom.

(6) Where a provision is rendered ineffective by subsection (1), the parties are free to agree other terms for payment.

In the absence of such agreement, the relevant provisions of the Scheme for Construction Contracts apply.

Supplementary provisions

114 The Scheme for Construction Contracts

(1) The Minister shall by regulations make a scheme ("the Scheme for Construction Contracts") containing provision about the matters referred to in the preceding provisions of this Part.

(2) Before making any regulations under this section the Minister shall consult such persons as he thinks fit.

(3) In this section "the Minister" means—
   (*a*) for England and Wales, the Secretary of State, and
   (*b*) for Scotland, the Lord Advocate.

(4) Where any provisions of the Scheme for Construction Contracts apply by virtue of this Part in default of contractual provision agreed by the parties, they have effect as implied terms of the contract concerned.

(5) Regulations under this section shall not be made unless a draft of them has been approved by resolution of each House of Parliament.

115 Service of notices, &c

(1) The parties are free to agree on the manner of service of any notice or other document required or authorised to be served in pursuance of the construction contract or for any of the purposes of this Part.

(2) If or to the extent that there is no such agreement the following provisions apply.

(3) A notice of other document may be served on a person by any effective means.

(4) If a notice or other document is addressed, pre-paid and delivered by post—
   (*a*) to the addressee's last known principal residence or, if he is or has been carrying on a trade, profession or business, his last known principal business address, or
   (*b*) where the addressee is a body corporate, to the body's registered or principal office,

   it shall be treated as effectively served.

(5) This section does not apply to the service of documents for the purposes of legal proceedings, for which provision is made by rules of court.

(6) References in this Part to a notice or other document include any form of communication in writing and references to service shall be construed accordingly.

116 Reckoning periods of time

(1) For the purposes of this Part periods of time shall be reckoned as follows.

(2) Where an act is required to be done within a specified period after or from a specified date, the period begins immediately after that date.

(3) Where the period would include Christmas Day, Good Friday or a day which under the Banking and Financial Dealings Act 1971 (1971 c. 80) is a bank holiday in England and Wales or, as the case may be, in Scotland, that day shall be excluded.

117 Crown application

(1) This Part applies to a construction contract entered into by or on behalf of the Crown otherwise than by or on behalf of Her Majesty in her private capacity.

(2) This Part applies to a construction contract entered into on behalf of the Duchy of Cornwall notwithstanding any Crown interest.

(3) Where a construction contract is entered into by or on behalf of Her Majesty in right of the Duchy of Lancaster, Her Majesty shall be represented, for the purposes of any adjudication or other proceedings arising out of the contract by virtue of this Part, by the Chancellor of the Duchy or such person as he may appoint.

(4) Where a construction contract is entered into on behalf of the Duchy of Cornwall, the Duke of Cornwall or the possessor for the time being of the Duchy shall be represented, for the purposes of any adjudication or other proceedings arising out of the contract by virtue of this Part, by such person as he may appoint.

PART III Architects

The Architects Registration Board

118 The Board and its committees

(1) The Architects' Registration Council of the United Kingdom established under the Architects (Registration) Act 1931 ("the

1931 Act"; 1931 c. 33) shall be known as the Architects Registration Board.

(2) The Board of Architectural Education, the Admission Committee and the Discipline Committee constituted under the 1931 Act are abolished.

(3) In section 3 of the 1931 Act (constitution and functions of Architects' Registration Council), after subsection (2) insert—

"(2A) Part I of the First Schedule to this Act makes provision about the constitution and proceedings of tbe Board."

Source: *The Housing Grants, Construction and Regeneration Act* 1996. HMSO, London, 1996.

Appendix B

Statutory Instrument 1998 No. 649

The Scheme for Construction Contracts (England and Wales) Regulations 1998

The text of this Internet version of the Statutory Instrument has been prepared to reflect the text as it was Made. The authoritative version is in printed form and is published by the Stationery Office Limited as the **The Scheme for Construction Contracts (England and Wales) Regulations (1998)**, ISBN 0 11 065709 8, £1.95 sterling. For details of how to obtain a printed copy see How to obtain the Stationery Office Limited titles.

Statutory Instruments

The Scheme for Construction Contracts (England and Wales) Regulations 1998)

Made 6th March 1998
Coming into force 1st May 1998

The Sectretary of State, in exercise of the powers conferred on him by section 108(6), 114 and 146(1) and (2) of the Housing Grants, Construction and Regeneration Act 1996(1), and of all other powers enabling him in that behalf, having consulted such persons as he thinks fit, and draft Regulations having been approved by both Houses of Parliament, herby makes the following Regulations:

Citation, commencement, extent and interpretation

1. (1) These Regulations may be cited as The Scheme for Construction Contracts (England and Wales) Regulations 1998 and shall come into force at the end of the period of 8 weeks beginning with the day on which it is made (the "commencement date").
   (2) These Regulations shall extend only to England and Wales.
   (3) In these Regulations, "the Act" means the Housing Grants, Construction and Regeneration Act 1996.

The Scheme for Construction Contracts

2. Where a construction contract does not comply with the requirements of section 108(1) to (4) of the Act, the adjudication provisions in Part 1 of the Schedule to these Regulations shall apply.
3. Where—
   (*a*) the parties to a construction contract are unable to reach agreement for the purposes mentioned respectively in sections 109, 111 and 113 of the Act, or
   (*b*) a construction contract does not make provision as required by section 110 of the Act,

the relevant provisions in Part II of the Schedule to these Regulations shall apply.

4. The provisions in the Schedule to these Regulations shall be the Scheme for Construction Contracts for the purposes of section 114 of the Act.

Signed by authority of the Secretary of State

Nick Raynsford
Parliamentary Under-Secretary of State,
Department of the Environment,
Transport and the Regions

6th March 1998

Part I. Adjudication

Notice of Intention to seek Adjudication

1. (1) Any party to a construction contract (the "referring party") may give written notice (the "notice of adjudication") of his intention to refer any dispute arising under the contact to adjudication.

(2) The notice of adjudication shall be given to every other party to the contract.

(3) The notice of adjudication shall set out briefly—

(*a*) the nature and a brief description of the dispute and of the parties involved

(*b*) details of where and when the dispute has arisen,

(*c*) the nature of the redress which is sought, and

(*d*) the names and addresses of the parties to the contract (including, where appropriate, the addresses which the parties have specified for the giving of notices).

2. (1) Following the giving of a notice of adjudication and subject to any agreement between the parties to the dispute as to who shall act as adjudicator—

(*a*) the referring party shall request the person (if any) specified in the contract to act as adjudicator, or

(*b*) if no person is named in the contract or the person named has already indicated that he is unwilling or unable to act, and the contract provides for a specified nominating body to select a person, the referring party shall request the nominating body named in the contract to select a person to act as adjudicator, or

(*c*) where neither paragraph (a) nor (b) above applies, or where the person referred to in (a) has already indicated that he is unwilling or unable to act and (b) does not apply, the referring party shall request an adjudicator nominating body to select a person to act as adjudicator.

(2) A person requested to act as adjudicator in accordance with the provisions of paragraph (1) shall indicate whether or not he is willing to act within two days of receiving the request.

(3) In this paragraph, and in paragraphs 5 and 6 below, an "adjudicator nominating body" shall mean a body (not being a natural person and not being a party to the dispute) which holds itself out publicly as a body which will select an adjudicator when requested to do so by a referring party

3. The request referred to in paragraphs 2.5 and 6 shall be accompanied by a copy of the notice of adjudication.

4. Any person requested or selected to act as adjudicator in accordance with paragraphs 2, 5 or 6 shall be a natural person acting in his personal capacity. A person requested or selected to act as an adjudicator shall not be an employee of any of the parties to the dispute and shall declare any interest, financial or otherwise, in any matter relating to the dispute.

5. (1) The nominating body referred to in paragraphs 2(1)(b) and 6(1)(b) or the adjudicator nominating body referred to in paragraphs 2(1)(c), 5(2)(b) and 6(1)(c) must communicate the selection of an adjudicator to the referring party within five days of receiving a request to do so.

(2) Where the nominating body or the adjudicator nominating body fails to comply with paragraph (1) the referring party may—

(*a*) agree with the other parties to the dispute to request a specified person to act as adjudicator, or (*b*) request any other adjudicator nominating body to select a person to act as adjudicator.

(3) The person requested to act as adjudicator in accordance with the provisions of paragraphs (1) or (2) shall indicate whether or not he is willing to act within two days of receiving the request.

6. (1) Where another adjudicator who is named in the contract indicates to the parties that he is unable or unwilling to act or where he fails to respond in accordance with paragraph 2(2), the referring party may

(*a*) request another person (if any) specified in the contract to act as adjudicator or

(*b*) request the nominating body (if any) referred to in the contract to select a person to act as adjudicator, or

(*c*) request any other adjudicator nominating body to select a person to act as adjudicator.

(2) The person requested to act in accordance with the provisions of paragraph (1) shall indicate whether or not he is willing to act within two days of receiving the request.

7. (1) Where an adjudicator has been selected in accordance with paragraphs 2, 5 or 6. the referring party shall, not later than seven days from the date of the notice of adjudication, refer the dispute in writing (the "referral notice") to the adjudicator.

(2) A referral notice shall be accompanied by copies of or relevant extracts from, the construction contract and such other documents as the referring party intends to rely upon.

(3) The referring party shall, at the same time as he sends to the adjudicator the documents referred to in paragraphs (1) and (2), send copies of those documents to every other party to the dispute.

8. (1) The adjudicator may, with the consent of all the parties to those disputes, adjudicate at the same time on one or more disputes under the same contract.

(2) The adjudicator may, with the consent of all the parties to those disputes, adjudicate at the same time on related disputes under different contracts, whether or not one or more of those parties is a party to those disputes.

(3) All the parties in paragraphs (1) and (2) respectively may agree to extend the period within which the adjudicator may reach a decision in relation to all or any of these disputes.

(4) Where an adjudicator ceases to act because a dispute is to be adjudicated on by another person in terms of this paragraph, that adjudicator's fees and expenses shall be determined in accordance with paragraph 25.

9. (1) An adjudicator may resign at any time on giving notice in writing to the parties to the dispute.

(2) An adjudicator must resign where the dispute is the same or substantially the same as one which has previously been referred to adjudication and a decision has been taken in that adjudication.

(3) Where an adjudicator ceases to act under paragraph 9(1) —

(*a*) the referring party may serve a fresh notice under paragraph 1 and shall request an adjudicator to act in accordance with paragraphs 2 to 7; and

(*b*) if requested by the new adjudicator and insofar as it is reasonably practicable, the parties shall supply him with copies of all documents which they had made available to the previous adjudicator.

(4) Where an adjudicator resigns in the circumstances referred to in paragraph (2), or where the dispute varies significantly from the dispute referred to him in the referral notice and for that reason he is not competent to decide it, the adjudicator shall be entitled to the payment of such reasonable amount as he may determine by way of fees and expenses reasonably incurred by him. The parties shall be jointly and severally liable for any sum which remains outstanding

following the making of any determination on how the payment shall be apportioned.

10. Where any party to the dispute objects to the appointment of a particular person as adjudicator that objection shall not invalidate the adjudicator's appointment nor any decision he may reach in accordance with paragraph 20.

11. (1) The parties to a dispute may at any time agree to revoke the appointment of the adjudicator. The adjudicator shall be entitled to the payment of such reasonable amount as he may determine by way of fees and expenses incurred by him. The parties shall be jointly and severally liable for any sum which remains outstanding following the making of any determination on how the payment shall be apportioned.

(2) Where the revocation of the appointment of the adjudicator is due to the default or misconduct of the adjudicator, the parties shall not be liable to pay the adjudicator's fees and expenses.

Powers of the adjudicator

12. The adjudicator shall—

(*a*) act impartially in carrying out his duties and shall do so in accordance with any relevant terms of the contract and shall reach his decision in accordance with the applicable law in relation to the contract; and

(*b*) avoid incurring unnecessary expense.

13. The adjudicator may take the initiative in ascertaining the facts and the law necessary to determine the dispute, and shall decide on the procedure to be followed in the adjudication. In particular he may—

(*a*) request any party to the contract to supply him with such documents as he may reasonably require including, if he so directs, any written statement from any party to the contract supporting or supplementing the referral notice and any other documents given under paragraphs 7(2),

(*b*) decide the language or languages to be used in the adjudication and whether a translation of any document is to be provided and if so by whom,
(*c*) meet and question any of the parties to the contract and their representatives,
(*d*) subject to obtaining any necessary consent from a third party or parties, make such site visits and inspections as he considers appropriate, whether accompanied by the parties or not,
(*e*) subject to obtaining any necessary consent from a third party or parties, carry out any tests or experiments,
(*f*) obtain and consider such representations and submissions as he requires, and, provided he has notified the parties of his intention, appoint experts, assessors or legal advisers,
(*g*) give directions as to the timetable for the adjudication any deadlines, or limits as to the length of written documents or oral representations to be complied with, and
(*h*) issue other directions relating to the conduct of the adjudication.

14. The parties shall comply with any request or direction of the adjudicator in relation to the adjudication.

15. If, without showing sufficient cause, a party fails to comply with any request, direction or timetable of the adjudicator made in accordance with his powers, fails to produce any document or written statement requested by the adjudicator or in any other way fails to comply with a requirement under these provisions relating to the adjudication, the adjudicator may—
(*a*) continue the adjudication in the absence of that party or of the document or written statement requested,
(*b*) draw such inferences from that failure to comply as circumstances may, in the adjudicator's opinion, be justified, and
(*c*) make a decision on the basis of the information before him attaching such weight as he thinks fit to any evidence submitted to him outside any period he may have requested or directed.

16. (1) Subject to any agreement between the parties to the contrary and to the terms of paragraph (2) below, any party to the dispute may be assisted by, or represented by, such advisers or representatives (whether legally qualified or not) as he considers appropriate.

(2) Where the adjudicator is considering oral evidence or representations, a party to the dispute may not be represented by more than one person, unless the adjudicator gives directions to the contrary.

17. The adjudicator shall consider any relevant information submitted to him by any of the parties to the dispute and shall make available to them any information to be taken into account in reaching his decision.

18. The adjudicator and any party to the dispute shall not disclose to any other person any information or document provided to him in connection with the adjudication which the party supplying it has indicated is to be treated as confidential except to the extent that it is necessary for the purposes of, or in connection with the adjudication

19. (1) The adjudicator shall reach his decision not later than—

(*a*) twenty eight days after the date of the referral notice mentioned in paragraph 7(1),

(*b*) forty two days after the date of the referral notice if the referring party so consents, or

(*c*) such period exceeding twenty eight days after the referral notice as the parties to the dispute may, after the giving of that notice, agree.

(2) Where the adjudicator fails, for any reason, to reach his decision in accordance with paragraph (1)

(*a*) any of the parties to the dispute may serve a fresh notice under paragraph 1 and shall request an adjudicator to act in accordance with paragraphs 2 to 7; and

(*b*) if requested by the new adjudicator and insofar as it is reasonably practicable, the parties shall supply him with copies of all documents which they had made available to the previous adjudicator.

(3) As soon as possible after he has reached a decision, the adjudicator shall deliver a copy of that decisions to each of the parties to the contract.

Adjudicator's decision

20. The adjudicator shall decide the matters in dispute. He may take into account any other matters which the parties to the dispute agree should be within the scope of the adjudication or which are matters under the contract which he considers are necessarily connected with the dispute. In particular, he may—

(*a*) open up, revise and review any decision taken or any certificate given by any person referred to in the contract unless the contract states that the decision or certificate is final and conclusive,

(*b*) decide that any of the parties to the dispute is liable to make a payment under the contract (whether in sterling or some other currency) and, subject to section 111(4) of the Act, when that payment is due and the final date for payment,

(*c*) having regard to any term of the contract relating to the payment of interest decide the circumstances in which, and the rates at which, and the periods for which simple or compound rates of interest shall be paid.

21. In the absence of any directions by the adjudicator relating to the time for performance of his decision, the parties shall be required to comply with any decision of the adjudicator immediately on delivery of the decision to the parties in accordance with this paragraph.

22. If requested by one of the parties to the dispute, the adjudicator shall provide reasons for his decision.

Effects of the decision

23. (1) In his decision, the adjudicator may, if he thinks fit, order any of the parties to comply peremptorily with his decision or any part of it.

(2) The decision of the adjudicator shall be binding on the parties, and they shall comply with it until the dispute is finally determined by legal proceedings, by arbitration (if the contract provides for arbitration or the parties otherwise agree to arbitration) or by agreement between the parties.

24. Section 42 of the Arbitration Act 1996 shall apply to this Scheme subject to the following modifications—

(*a*) in subsection (2) for the word "tribunal" wherever it appears there shall be substituted the word "adjudicator",

(*b*) in subparagraph (b) of subsection (2) for the words "arbitral proceedings" there shall be substituted the word "adjudication",

(*c*) subparagraph (c) of subsection (2) shall be deleted, and

(*d*) subsection (3) shall be deleted.

25. The adjudicator shall be entitled to the payment of such reasonable amount as he may determine by way of fees and expenses reasonably incurred by him. The parties shall be jointly and severally liable to par any sum which remains outstanding following the making of any determination on how the payment shall be apportioned.

26. The adjudicator shall not be liable for anything done or omitted in the discharge or purported discharge of his functions as adjudicator unless the act or omission is in bad faith and any employee or agent of the adjudicator shall be similarly protected from liability.

Part II. Payment

Entitlement to and amount of stage payments

1. Where the parties to a relevant construction contract fail to agree—

(*a*) the amount of any instalment or stage or periodic payment for any work under the contract, or (*b*) the intervals at which, or circumstances in which, such payments become due under that contract, or

(*c*) both of the matters mentioned in sub-paragraphs (a) and (b) above, the relevant provisions of paragraphs 2 to 4 below shall apply.

2. (1) The amount of any payment by way of instalments or stage or periodic payments in respect of a relevant period shall be the difference between the amount determined in accordance with sub-paragraph (2) and the amount determined in accordance with sub-paragraph (3).

(2) The aggregate of the following amounts—

(*a*) an amount equal to the value of any work performed in accordance with the relevant construction contract during the period from the commencement of the contract to the end of the relevant period (excluding any amount calculated in accordance with sub-paragraph (b)),

(*b*) where the contract provides for payment for materials, an amount equal to the value of any materials manufactured on site or brought onto site for the purposes of the works during the period from the commencement of the contract to the end of the relevant period, and

(*c*) any other amount or sum which the contract specifies shall be payable during or in respect of the period from the commencement of the contract to the end of the relevant period.

(3) The aggregate of any sums which have been paid or are due for payment by way of instalments, stage or periodic payments during the period from the commencement of the contract to the end of the relevant period.

(4) An amount calculated in accordance with this paragraph shall not exceed the difference between—

(*a*) the contract price, and

(*b*) the aggregate of the instalments or stage or periodic payments which have become due.

Dates for payment

3. Where the parties to a construction contract fail to provide an adequate mechanism for determining either what payments become due under the contract. or when they become due for payment, or both, the relevant provisions of paragraphs 4 to 7 shall apply.
4. Any payment of a kind mentioned in paragraph 2 above shall become due on whichever of the following dates occurs later—
   (*a*) the expiry of 7 days following the relevant period mentioned in paragraph 2(1) above, or
   (*b*) the making of a claim by the payee.
5. The final payment payable under a relevant construction contract, namely the payment of an amount equal to the difference (if any) between—
   (*a*) the contract price, and
   (*b*) the aggregate of any instalment or stage or periodic payments which have becoming due under the contract,

   shall become due on the expiry of—
   (*a*) 30 days following completion of the work, or
   (*b*) the making of a claim by the payee,

   whichever is the later.
6. Payment of the contract price under a construction contract (not being a relevant construction contract) shall become due on
   (*a*) the expiry of 30 days following the completion of the work, or
   (*b*) the making of a claim by the payee,

   whichever is the later.
7. Any other payment under a construction contract shall become due
   (*a*) on the expiry of 7 days following the completion of the work to which the payment relates, or
   (*b*) the making of a claim by the payee,

   whichever is the later.

Final date for payment

8. (1) Where the parties to a construction contract fail to provide a final date for payment in relation to any sum which becomes due under a construction contract, the provisions of this paragraph shall apply.

(2) The final date for the making of any payment of a kind mentioned in paragraphs 2, 5, 6 or 7, shall be 17 days from the date that payment becomes due.

Notice specifying amount of payment

9. A party to a construction contract shall not later than 5 days after the date on which any payment

(*a*) becomes due from him, or

(*b*) would have become due, if—

(*i*) the other party had carried out his obligations under the contract, and

(*ii*) no set-off or abatement was permitted by reference to any sum claimed to be due under one or more other contracts,

give notice to the other party to the contract specifying the amount (if any) of the payment he has made or proposes to make, specifying to what the payment relates and the basis on which that amount is calculated.

Notice of intention to withhold payment

10. Any notice of intention to withhold payment mentioned in section 111 of the Act shall be given not later than the prescribed period, which is to say not later than 7 days before the final date for payment determined either in accordance with the construction contract, or where no such provision is made in the contract, in accordance with paragraph 8 above.

Prohibition of conditional payment provisions

11. Where a provision making payment under a construction contract conditional on the payer receiving payment from a third person is ineffective as mentioned in section 113 of the Act, and the parties have not agreed other terms for payment, the relevant provisions of—

(*a*) paragraphs 2, 4, 5, 7, 8, 9 and 10 shall apply in the case of a relevant construction contract, and

(*b*) paragraphs 6, 7, 8, 9 and 10 shall apply in the case of any other construction contract.

Interpretation

12. In this Part of the Scheme for Construction Contracts—

"claim by the payee" means a written notice given by the party carrying out work under a construction contract to the other party specifying the amount of any payment or payments which he considers to be due and the basis on which it is, or they are calculated;

"contract price" means the entire sum payable under the construction contract in respect of the work;

"relevant construction contract" means any construction contract other than one—

(*a*) which specifies that the duration of the work is to be less than 45 days, or

(*b*) in respect of which the parties agree that the duration of the work is estimated to be less than 45 days;

"relevant period" means a period which is specified in, or is calculated by reference to the construction contract or where no such period is so specified or is so calculable, a period of 28 days;

"value of work" means an amount determined in accordance with the construction contract under which the work is performed or where the contract contains no such provision, the cost of any work performed in accordance with that contract together with an amount equal to any overhead on profit included in the contract price:

"work" means any of the work or services mentioned in section 104 of the Act.

Explanatory Note

(This note is not part of the Order)

Part II of the Housing Grants, Construction and Regeneration Act 1996 makes provision in relation to construction contracts. Section 114 empowers the Secretary of State to make the Scheme for Construction Contracts. Where a construction contract does not comply with the requirements of sections 108 to 111 (adjudication of disputes and payment provisions), and section 113 (prohibition of conditional payment provisions), the relevant provisions of the scheme for Construction Contracts have effect.

The Scheme which is contained in the Schedule to these Regulations is in two parts. Part I provides for the selection and appointment of an adjudicator, gives powers to the adjudicator to gather and consider informaton, and makes provisions in respect of his decisions. Part II makes provision with respect to payments under a construction contract where either the contract fails to make provision or the parties fail to agree—

(*a*) the method for calculating the amout of any instalment, stage or periodic payment,

(*b*) the due date and the final date for payments to be made, and

(*c*) prescribes the period within which a notice of intention to withhold payment must be given.

Appendix C

Differences between the versions of the Scheme for England and Wales and the Scheme for Scotland

PART I. Adjudication

Paragraph	Comments
9(4)	For England and Wales the payment terms for the adjudicator in the event of different situations arising have been considered separately for each event leading to repetition of the same words whereas in the version for Scotland they have been consolidated and cross-referenced to paragraph 25. The Scottish version is to be preferred.
11(1)	Same comment.
15(b)	Better drafting of the Scottish version using the same words.
20(1)	The Scottish version reads: The adjudicator shall decide the matters in dispute *and may make a decision on different aspects of the dispute at different times.* The words in *italics* do not appear in the version for England and Wales. They are a useful addition and will help the adjudicator to reach an earlier decision on some of the issues than is likely under the English version. In England and Wales adjudicators are likely to wait until near the end of the period allowed for making

the decision in case further information comes in. Under the Scottish version the adjudicator can make decisions on preliminary issues immediately if he thinks fit. The Scottish version is to be preferred.

21 The Scottish version corrects an obvious defect in the drafting of the English version. The adjudicator is required to deliver his decision to the parties as soon as possible after reaching his decision under clause 19. The English version refers to delivery under this paragraph (i.e. 21) when it should be 19.

24 This is the crucial difference between the two versions resulting from the different legal positions in respect of arbitration. S108 (6) of the Act states:

> *For England and Wales, the Scheme may apply the provisions of the Arbitration Act 1996 with such adaptations and modifications as appear to the Minister making the scheme to be appropriate.*
>
> *For Scotland, the Scheme may include provision conferring powers on courts in relation to adjudication and provision relating to the enforcement of the adjudicator's decision.*

The differences between the two versions therefore follow from the Act. The Scottish version states:

> *Where a party or the adjudicator wishes to register the decision for execution in the Books of Council and Session, any other party shall, on being requested to do so, forthwith consent to such registration by subscribing the decision before a witness.*

Adjudicators in Scotland are likely, as a routine, to require one party to register the decision. If they do not do this routinely they will leave themselves open to accusations of bias on the basis of why require me to do it when you did not require someone else to?

Presumably the party required to register the decision will be the winning side, and that will leave the other party to decide whether or not to apply to the courts for an interdict when they are unwilling to comply with the decision.

It must be expected that in Scotland, as in England, the decisions of adjudicators in the smaller disputes will be acted upon but when large sums of money are involved and a party is unwilling to accept the decision that party will take every legal step that is available to prevent the decision being enforced.

That is a failing of the Act not the Scheme.

25 This paragraph has been redrafted in the Scottish version and is much clearer than the English version. The English version has consistently misunderstood the meaning of 'joint and several' liability. There is no difference in substance between the two versions, merely that the Scottish version is clearer and more accurately drafted. Again the Scottish version is to be preferred.

PART II. Payment

5(i) The only change in the Part is to this paragraph and it picks up an obvious drafting error in the English version. It has no significance. The Scottish version is to be preferred.

Statutory Instrument 1998 No.687 (S. 34)

The Scheme for Construction Contracts (Scotland) Regulations 1998

The text of this Internet version of the Statutory Instruments has been prepared to reflect the text as it was Made. The authoritative version is the Queen's Printer copy published by The Stationery Office Limited as the *The Scheme for Construction Contracts (Scotland) Regulation* 1998, ISBN 0 11 055746 8. Purchase this item. For details of how to obtain an official copy see How to obtain The Stationery Office Limited titles.

To ensure fast access over slow connections, large documents have been segmented into "chunks". Where you see a "continue" button at the bottom of the page of text, this indicates that there is another chunk of text available.

Statutory Instruments 1998 No. 687 (S. 34)

The scheme for Construction Contracts (Scotland) Regulations 1998

Made	*6th March 1998*
Coming into force	*1st May 1998*

The Lord Advocate, in exercise of the powers conferred on him by sections 108(6), 114 and 146 of the Housing Grants, Construction and Regeneration Act 1996 [1] and of all other powers enabling him in that behalf, having consulted such persons as he thinks fit, hereby makes the following Regulations, a draft of which has been laid before and has been approved by resolution of each House of Parliament:

Citation, commencement and extent

1. (1) These Regulations may be cited as the Scheme for Construction Contracts (Scotland) Regulations 1998 and shall come into force at the end of the period of 8 weeks beginning with the day on which they are made.

   (2) These Regulations extend to Scotland only.

Interpretation

2. In these Regulations, "the Act" means the Housing Grants, Construction and Regeneration Act 1996.

The Scheme for Construction Contracts (Scotland)

3. Where a construction contract does not comply with the requirements of subsections(1) to (4) of section 108 of the Act, the adjudication provisions in PartI of the Schedule to these Regulations shall apply.

4. Where—

(*a*) the parties to a construction contract are unable to reach agreement for the purposes mentioned respectively in sections 109, 111 and 113 of the Act; or

(*b*) a construction contract does not make provision as required by section 110 of the Act.

the relevant provisions in Part II of the Schedule to these Regulations shall apply.

5. The provisions in the Schedule to these Regulations shall be the Scheme for Construction Contracts (Scotland) for the purposes of section 114 of the Act.

Hardie
Lord Advocate
Edinburgh
6th March 1998

Schedule

Regulations 3 to 5

The Scheme for Construction Contracts (Scotland) Regulations 1998

The text of the above is reproduced below.

Part I. Adjudication

Notice of intention to seek adjudication

1. (1) Any party to a construction contract ("the referring party") may give written notice ("the notice of adjudication") of his intention to refer any dispute arising under the contract to adjudication.
   (2) The notice of adjudication shall be given to every other party to the contract.
   (3) The notice of adjudication shall set out briefly—
      (*a*) the nature and a brief description of the dispute and of the parties involved;
      (*b*) details of where and when the dispute has arisen;
      (*c*) the nature of the redress which is sought; and
      (*d*) the names and addresses of the parties to the contract (including, where appropriate, the addresses which the parties have specified for the giving of notices).
2. (1) Following the giving of a notice of adjudication and subject to any agreement between the parties to the dispute as to who shall act as adjudicator—
      (*a*) the referring party shall request the person (if any) specified in the contract to act as adjudicator;
      (*b*) if no person is named in the contract or the person named has already indicated that he is unwilling or unable to act, and the contract provides for a specified nominating body to select a person, the referring party shall request the nominating body named in the contract to select a person to act as adjudicator; or
      (*c*) where neither head (*a*) nor (*b*) above applies, or where the person referred to in (*a*) has already indicated that he is unwilling or unable to act and (*b*) does not apply, the referring party shall request an adjudicator nominating body to select a person to act as adjudicator.
   (2) A person requested to act as adjudicator in accordance with the provisions of sub-paragraph (1) shall indicate whether

or not he is willing to act within two days of receiving the request.

(3) In this paragraph, and in paragraphs 5 and 6 below, "an adjudicator nominating body" shall mean a body (not being a natural person and not being a party to the dispute) which holds itself out publicly as a body which will select an adjudicator when requested to do so by a referring party.

3. The request referred to in paragraphs 2, 5 and 6 shall be accompanied by a copy of the notice of adjudication.

4. Any person requested or selected to act as adjudicator in accordance with paragraphs 2, 5 or 6 shall be a natural person acting in his personal capacity. A person requested or selected to act as an adjudicator shall not be an employee of any of the parties to the dispute and shall declare any interest, financial or otherwise, in any matter relating to the dispute.

5. (1) The nominating body referred to in paragraphs 2(1)(b) and 6(1)(b) or the adjudicator nominating body referred to in paragraphs 2(1)(c), 5(2)(b) and 6(1)(c) must communicate the selection of an adjudicator to the referring party within five days of receiving a request to do so.

(2) Where the nominating body or the adjudicator nominating body fails to comply with sub-paragraph (1), the referring party may—

(*a*) agree with the other party to the dispute to request a specified person to act as adjudicator; or

(*b*) request any other adjudicator nominating body to select a person to act as adjudicator.

(3) The person requested to act as adjudicator in accordance with the provisions of sub-paragraph (1) or (2) shall indicate whether or not he is willing to act within two days of receiving the request.

6. (1) Where an adjudicator who is named in the contract indicates to the parties that he is unable or unwilling to act, or where he fails to respond in accordance with paragraph 2(2), the referring party may—

(*a*) request another person (if any) specified in the contract to act as adjudicator;
(*b*) request the nominating body (if any) referred to in the contract to select a person to act as adjudicator; or
(*c*) request any other adjudicator nominating body to select a person to act as adjudicator.

(2) The person requested to act in accordance with the provisions of sub-paragraph (1) shall indicate whether or not he is willing to act within two days of receiving the request.

7. (1) Where an adjudicator has been selected in accordance with paragraphs 2, 5 or 6, the referring party shall, not later than seven days from the date of the notice of adjudication, refer the dispute in writing ("the referral notice") to the adjudicator.

(2) A referral notice shall be accompanied by copies of, or relevant extracts from, the construction contract and such other documents as the referring party intends to rely upon.

(3) The referring party shall, at the same time as he sends to the adjudicator the documents referred to in sub-paragraphs (1) and (2), send copies of those documents to every other party to the dispute.

8. (1) The adjudicator may, with the consent of all the parties to those disputes, adjudicate at the same time on more than one dispute under the same contract.

(2) The adjudicator may, with the consent of all the parties to those disputes, adjudicate at the same time on related disputes under different contracts, whether or not one or more of those parties is a party to those disputes.

(3) All the parties in sub-paragraphs (1) and (2) respectively may agree to extend the period within which the adjudicator may reach a decision in relation to all or any of these disputes.

(4) Where an adjudicator ceases to act because a dispute is to be adjudicated on by another person in terms of this

paragraph, that adjudicator's fees and expenses shall be determined and payable in accordance with paragraph 25.

9. (1) An adjudicator may resign at any time on giving notice in writing to the parties to the dispute.

(2) An adjudicator must resign where the dispute is the same or substantially the same as one which has previously been referred to adjudication, and a decision has been taken in that adjudication.

(3) Where an adjudicator ceases to act under sub-paragraph (1)—

(*a*) the referring party may serve a fresh notice under paragraph 1 and shall request an adjudicator to act in accordance with paragraphs 2 to 7; and

(*b*) if requested by the new adjudicator and insofar as it is reasonably practicable, the parties shall supply him with copies of all documents which they had made available to the previous adjudicator.

(4) Where an adjudicator resigns in the circumstances mentioned in sub-paragraph (2), or where a dispute varies significantly from the dispute referred to him and for that reason he is not competent to decide it, that adjudicator's fees and expenses shall be determined and payable in accordance with paragraph 25.

10. Where any party to the dispute objects to the appointment of a particular person as adjudicator, that objection shall not invalidate the adjudicator's appointment nor any decision he may reach in accordance with paragraph 20.

11. (1) The parties to a dispute may at any time agree to revoke the appointment of the adjudicator and in such circumstances the fees and expenses of that adjudicator shall, subject to sub-paragraph (2), be determined and payable in accordance with paragraph 25.

(2) Where the revocation of the appointment of the adjudicator is due to the default or misconduct of the adjudicator,

the parties shall not be liable to pay the adjudicator's fees and expenses.

Powers of the adjudicator

12. The adjudicator shall—

(*a*) act impartially in carrying out his duties and shall do so in accordance with any relevant terms of the contract and shall reach his decision in accordance with the applicable law in relation to the contract; and

(*b*) avoid incurring unnecessary expense.

13. The adjudicator may take the initiative in ascertaining the facts and the law necessary to determine the dispute, and shall decide on the procedure to be followed in the adjudication. In particular, he may—

(*a*) request any party to the contract to supply him with such documents as he may reasonably require including, if he so directs, any written statement from any party to the contract supporting or supplementing the referral notice and any other documents given under paragraph 7(2);

(*b*) decide the language or languages to be used in the adjudication and whether a translation of any document is to be provided and, if so, by whom;

(*c*) meet and question any of the parties to the contract and their representatives;

(*d*) subject to obtaining any necessary consent from a third party or parties, make such site visits and inspections as he considers appropriate, whether accompanied by the parties or not;

(*e*) subject to obtaining any necessary consent from a third party or parties, carry out any tests or experiments;

(*f*) obtain and consider such representations and submissions as he requires, and, provided he has notified the parties of his intention, appoint experts, assessors or legal advisers;

(*g*) give directions as to the timetable for the adjudication, any deadlines, or limits as to the length of written documents or oral representations to be complied with; and

(*h*) issue other directions relating to the conduct of the adjudication.

14. The parties shall comply with any request or direction of the adjudicator in relation to the adjudication.

15. If, without showing sufficient cause, a party fails to comply with any request, direction or timetable of the adjudicator made in accordance with his powers, fails to produce any document or written statement requested by the adjudicator, or in any other way fails to comply with a requirement under these provisions relating to the adjudication, the adjudicator may—

(*a*) continue the adjudication in the absence of that party or of the document or written statement requested;

(*b*) draw such inferences from that failure to comply as may, in the adjudicator's opinion, be justified in the circumstances; and

(*c*) make a decision on the basis of the information before him, attaching such weight as he thinks fit to any evidence submitted to him outside any period he may have requested or directed.

16. (1) Subject to any agreement between the parties to the contrary and to the terms of sub-paragraph (2), any party to the dispute may be assisted by, or represented by, such advisers or representatives (whether legally qualified or not) as he considers appropriate.

(2) Where the adjudicator is considering oral evidence or representations, a party to the dispute may not be represented by more than one person, unless the adjudicator gives directions to the contrary.

17. The adjudicator shall consider any relevant information submitted to him by any of the parties to the dispute and shall make available to them any information to be taken into account in reaching his decision.

18. The adjudicator and any party to the dispute shall not disclose to any other person any information or document provided to him in connection with the adjudication which the party supplying it has indicated is to be treated as confidential, except to the extent that it is necessary for the purposes of, or in connection with, the adjudication.

19. (1) The adjudicator shall reach his decision not later than—

(*a*) twenty eight days after the date of the referral notice mentioned in paragraph 7(1);

(*b*) forty two days after the date of the referral notice if the referring party so consents; or

(*c*) such period exceeding twenty eight days after the referral notice as the parties to the dispute may, after the giving of that notice, agree.

(2) Where the adjudicator fails, for any reason, to reach his decision in accordance with sub-paragraph (1)—

(*a*) any of the parties to the dispute may serve a fresh notice under paragraph 1 and shall request an adjudicator to act in accordance with paragraphs 2 to 7; and

(*b*) if requested by the new adjudicator and insofar as it is reasonably practicable, the parties shall supply him with copies of all documents which they had made available to the previous adjudicator.

(*c*) As soon as possible after he has reached a decision, the adjudicator shall deliver a copy of that decision to each of the parties to the contract.

Adjudicator's decision

20. (1) The adjudicator shall decide the matters in dispute and may make a decision on different aspects of the dispute at different times.

(2) The adjudicator may take into account any other matters which the parties to the dispute agree should be within the scope of the adjudication or which are matters under the

contract which he considers are necessarily connected with the dispute and, in particular, he may—

(*a*) open up, review and revise any decision taken or any certificate given by any person referred to in the contract, unless the contract states that the decision or certificate is final and conclusive;

(*b*) decide that any of the parties to the dispute is liable to make a payment under the contract (whether in sterling or some other currency) and, subject to section 111(4) of the Act, when that payment is due and the final date for payment;

(*c*) having regard to any term of the contract relating to the payment of interest, decide the circumstances in which, the rates at which, and the periods for which simple or compound rates of interest shall be paid.

21. In the absence of any directions by the adjudicator relating to the time for performance of his decision, the parties shall be required to comply with any decision of the adjudicator immediately on delivery of the decision to the parties in accordance with paragraph 19(3).

22. If requested by one of the parties to the dispute, the adjudicator shall provide reasons for his decision.

Effects of the decision

23. (1) In his decision, the adjudicator may, if he thinks fit, order any of the parties to comply peremptorily with his decision or any part of it.

(2) The decision of the adjudicator shall be binding on the parties, and they shall comply with it, until the dispute is finally determined by legal proceedings, by arbitration (if the contract provides for arbitration or the parties otherwise agree to arbitration) or by agreement between the parties.

24. Where a party or the adjudicator wishes to register the decision for execution in the Books of Council and Session, any other party shall, on being requested to do so, forthwith consent to such registration by subscribing the decision before a witness.

25. (1) The adjudicator shall be entitled to the payment of such reasonable amount as he may determine by way of fees and expenses incurred by him and the parties shall be jointly and severally liable to pay that amount to the adjudicator.

(2) Without prejudice to the right of the adjudicator to effect recovery from any party in accordance with sub-paragraph (1), the adjudicator may by direction determine the apportionment between the parties of liability for his fees and expenses.

26. The adjudicator shall not be liable for anything done or omitted in the discharge or purported discharge of his functions as adjudicator unless the act or omission is in bad faith, and any employee or agent of the adjudicator shall be similarly protected from liability.

Part II. Payment

Entitlement to and amount of stage payments

1. Where the parties to a relevant construction contract fail to agree—

(*a*) the amount of any instalment or stage or periodic payment for any work under the contract;

(*b*) the intervals at which, or circumstances in which, such payments become due under that contract; or

(*c*) both of the matters mentioned in sub-paragraphs (*a*) and (*b*),

the relevant provisions of paragraphs 2 to 4 shall apply.

2. (1) The amount of any payment by way of instalments or stage or periodic payments in respect of a relevant period shall be the difference between the amount determined in accordance with sub-paragraph (2) and the amount determined in accordance with sub-paragraph (3).

(2) The aggregate of the following amounts—
  (*a*) an amount equal to the value of any work performed in accordance with the relevant construction contract during the period from the commencement of the contract to the end of the relevant period (excluding any amount calculated in accordance with head (*b*));
  (*b*) where the contract provides for payment for materials, an amount equal to the value of any materials manufactured on site or brought onto site for the purposes of the works during the period from the commencement of the contract to the end of the relevant period; and
  (*c*) any other amount or sum which the contract specifies shall be payable during or in respect of the period from the commencement of the contract to the end of the relevant period.

(3) The aggregate of any sums which have been paid or are due for payment by way of instalments, stage or periodic payments during the period from the commencement of the contract to the end of the relevant period.

(4) An amount calculated in accordance with this paragraph shall not exceed the difference between—
  (*a*) the contract price; and
  (*b*) the aggregate of the instalments or stage or periodic payments which have become due.

Dates for payment

3. Where the parties to a construction contract fail to provide an adequate mechanism for determining either what payments become due under the contract, or when they become due for payment, or both, the relevant provisions of paragraphs 4 to 7 shall apply.

4. Any payment of a kind mentioned in paragraph 2 above shall become due on whichever of the following dates occurs later—
  (*a*) the expiry of seven days following the relevant period mentioned in paragraph 2(1); or
  (*b*) the making of a claim by the payee.

5. The final payment payable under a relevant construction contract, namely the payment of an amount equal to the difference (if any) between—
   (*a*) the contract price; and
   (*b*) the aggregate of any instalment or stage or periodic payments which have become due under the contract,
   shall become due on—
      (*i*) the expiry of thirty days following completion of the work; or
      (*ii*) the making of a claim by the payee,
   whichever is the later.
6. Payment of the contract price under a construction contract (not being a relevant construction contract) shall become due on—
   (*a*) the expiry of thirty days following the completion of the work; or
   (*b*) the making of a claim by the payee,
   whichever is the later.
7. Any other payment under a construction contract shall become due on—
   (*a*) the expiry of seven days following the completion of the work to which the payment relates; or
   (*b*) the making of a claim by the payee,
   whichever is the later.

Final date for payment

8. (1) Where the parties to a contract fail to provide a final date for payment in relation to any sum which becomes due under a construction contract, the provisions of this paragraph shall apply.
   (2) The final date for the making of any payment of a kind mentioned in paragraphs 2, 5, 6 or 7 shall be 17 days from the date that payment becomes due.

Notice specifying amount of payment

9. A party to a construction contract shall, not later than 5 days after the date on which any payment—

(*a*) becomes due from him; or

(*b*) would have become due, if—

(*i*) the other party had carried out his obligations under the contract; and

(*ii*) no set-off or abatement was permitted by reference to any sum claimed to be due under one or more other contracts,

give notice to the other party to the contract specifying the amount (if any) of the payment he has made or proposes to make, specifying to what the payment relates and the basis on which that amount is calculated.

Notice of intention to withhold payment

10. Any notice of intention to withhold payment mentioned in section 111 of the Act shall be given not later than the prescribed period, which is to say not later than seven days before the final date for payment determined either in accordance with the construction contract or, where no such provision is made in the contract, in accordance with paragraph 8.

Prohibition of conditional payment provisions

11. Where a provision making payment under a construction contract conditional on the payer receiving payment from a third person is ineffective as mentioned in section 113 of the Act and the parties have not agreed other terms for payment, the relevant provisions of—

(*a*) paragraphs 2, 4, 5 and 7 to 10 shall apply in the case of a relevant construction contract; and

(*b*) paragraphs 6 to 10 shall apply in the case of any other construction contract.

Interpretation

12. In this Part—

"claim by the payee" means a written notice given by the party carrying out work under a construction contract to the other party specifying the amount of any payment or payments which he considers to be due, specifying to what the payment relates (or payments relate) and the basis on which it is, or they are, calculated;

"contract price" means the entire sum payable under the construction contract in respect of the work;

"relevant construction contract" means any construction contract other than one—

(*a*) which specifies that the duration of the work is to be less than 45 days; or

(*b*) in respect of which the parties agree that the duration of the work is estimated to be less than 45 days;

"relevant period" means a period which is specified in, or is calculated by reference to, the construction contract or, where no such period is so specified or is so calculable, a period of 28 days;

"value of work" means an amount determined in accordance with the construction contract under which the work is performed or, where the contract contains no such provision, the cost of any work performed in accordance with that contract together with an amount equal to any overhead or profit included in the contract price;

"work" means any of the work or services mentioned in section 104 of the Act.

Appendix D

A Case Study

Documents:

(*a*) Notice of Adjudication.
(*b*) Application for the selection of an adjudicator to an ANB.
(c) Typical ANB Application Form (The Institution of Civil Engineers).
(*d*) The Referral Notice (including supporting documents A, B and C).
(*e*) The Receiving Party's Statement (including supporting document).
(*f*) The Referring Party's reply to the Receiving Party's Statement.
(*g*) The Decision.

SuperConstruct Limited

Telephone: 01245 100100
Facsimile: 01245 100101

Unit 4
Bene Volent Trading Estate
Greenlands
Surrey OR7 8SA

Mr A Grade,
EasyView Cable Co. Ltd.,
Satellite House,
Great Sight Street,
London, E43A 2IC.

20 July

Dear Mr Grade,

Re: Newtown Franchise Area - Contract No.4

NOTICE OF ADJUDICATION.

Following a number of meetings with your Project Manager, Mr S.O. Tight, we have failed to reach agreement on two matters concerning the interpretation of the contract. We now give formal notice that we wish these matters to be referred to adjudication.

Since there is no provision for adjudication in our contract the adjudication will have to be conducted in accordance with the Scheme for Construction Contracts published by the Government.

Under Paragraph 1(3) of the Scheme we are required to give you the following information:

Item (a)

(i) The dispute involves the interpretation of Item 15 of the Schedule of Rates contained in the contract. The item reads "Backfill trench with imported material as directed" and the dispute is whether or not our quoted price includes the supply of the imported materials, and

(ii) Since the contract does not define the width of the trench how do we calculate the quantity of imported materials?

(iii) The parties involved in the dispute are you and us only.

Item (b)

The dispute has arisen because we were unable to resolve with your Project Manager, Mr Samuel Tight, these two matters at a meeting held on 16 June and he has subsequently refused to discuss the matter further.

Item (c)

We will be asking the adjudicator for a decision that the item does not include the supply of the materials, and the width of the trench (and other items) should be measured in accordance with the Civil Engineering Standard Method of Measurement (CESMM3) in the absence of any method of measurement in the contract.

Item (d)

The names and addresses of the parties are as given at the top of this letter.

In the absence of a nominating body for adjudicators in our contract we are applying today to the Institution of Civil Engineers, who we know act as an appointing body, for a nomination.

Yours faithfully,
For and on behalf of SuperConstruct Ltd,

I.B.Noah
Director

SuperConstruct Limited

Telephone: 01245 100100
Facsimile: 01245 100101

Unit 4
Bene Volent Trading Estate
Greenlands
Surrey OR7 8SA

Dispute Administration Service
The Institution of Civil Engineers,
1 Great George Street,
Westminster,
London SW1P 3AA.

20th July

Dear Sir,

Re: Appointment of an Adjudicator
Newtown Franchise Area - Contract No.4

We have a dispute with the employer, the EasyView Cable Co. Limited, on this contract and we wish to have an adjudicator appointed. The adjudication will be under the Scheme for Construction Contracts.

We enclose:
Your application form duly completed, and
(a) Our Notice of Dispute sent to EasyView today.
(b) Our cheque for the appointment fee of £150.

We look forward to receiving the name of the appointed person so that we can send our referral notice to him as early as possible.

Yours faithfully,
For and on behalf of SuperConstruct Ltd.

I.B.Noah,
Director

APPLICATION FOR THE SELECTION/APPOINTMENT OF AN ADJUDICATOR

Dispute Administration Service
The Institution of Civil Engineers
One Great George Street
Westminster
London SW1P 3AA

Date:.... 20th July............................

The dispute and/or difference described in the attached Notice of Adjudication Reference AB/SL has arisen and since an Adjudicator has not been agreed between the Parties I/we hereby apply to you to select/appoint an Adjudicator.

I/we agree to meet all the reasonable costs incurred by the person selected/appointed by you if I/we am/are not entitled to make this application in accordance with the agreement between the parties.

I/we enclose a cheque for £150.00 + VAT in respect of the charge made by the Institution towards the administrative cost in connection with this application.

Yours faithfully,

For and on behalf of :
Name: A.Brain

Address: SuperConstruct Limited,
Unit 4,
Bene Volent Trading Estate,
Greenlands,
Surrey OR7 8SA.

NOTES

1. The Institution will make a selection/appointment upon the application of any person using this form. The validity of the application will not be investigated.

2. In making this application the Applicant undertakes to meet the reasonable charges of the person selected/appointed by the Institution pursuant to this application should the adjudication not proceed.

3. If the validity of the application is challenged then the person appointed and the parties involved must resolve the challenge.

4. Neither the Institution nor its servants or agents nor its members shall be liable to any party for any act, omission or misconduct in connection with any selection or appointment made or any adjudication conducted in consequence of this application.

NOTICE OF ADJUDICATION AND FEE MUST ACCOMPANY THIS FORM

SELECTION/APPOINTMENT OF AN ADJUDICATOR BY THE INSTITUTION OF CIVIL ENGINEERS

To (Name and Address of Applicant):
SuperConstruct Limited,
Unit 4,
Bene Volent Trading Estate,
Greenlands,
Surrey OR7 8SA.

I hereby select/appoint:
Mr A. Wiseman

of
2 Oracle Way,
Crawley,
Sussex.

Adjudicator for the dispute or difference described in the Notice of Adjudication attached to the application with reference

Dated: 25 July

E Stanton

E.Stanton (Mrs)
On behalf of The Institution of Civil Engineers

Copies for information:
The Adjudicator
The Parties (other than the Applicant)
EasyView Cable Co Ltd.,
Satellite House,
Great Sight Street,
London, E43A 2IC.

SuperConstruct Limited

Telephone: 01245 100100
Facsimile: 01245 100101

Unit 4
Bene Volent Trading Estate
Greenlands
Surrey OR7 8SA

Mr A. Wiseman,
2 Oracle Road,
Crawley,
Sussex.

27 July

Dear Sir,

Adjudication between SuperConstruct Limited and EasyView Cable Co Ltd
Newtown Franchise Area Contract No 3.

REFERRAL NOTICE

We have been notified by the Institution of Civil Engineers that you have been selected as the adjudicator for this dispute and confirm our telephone conversation with you in which you accepted the appointment and said that you were willing and able to accept the appointment.

We now attach as part of this Referral Notice our statement and a copy of the Notice of Adjudication for your information.

Yours faithfully,

I.B.Noah,
Director
SuperConstruct Limited

cc EasyView Cable Company Ltd.

REFERRAL NOTICE

Introduction

SuperConstruct [SuperConstruct] entered into a contract with EasyView Cable Co Ltd [EasyView] for the installation of cable ducts in The Newtown Franchise Area No 3 on 13 February in the sum of £3,823,207.34.

The works commenced on 23 February and a number of differences on the interpretation of the contract immediately became apparent. A number of meetings were held between representatives of SuperConstruct and Mr Tight, the Contract Administrator, at which no agreement was reached and it became apparent that no agreement could be reached.

SuperConstruct therefore on 20 July gave Notice of Adjudication under the Housing Grants Construction and Regeneration Act 1996 because there was no adjudication provision in the contract. At the same time SuperConstruct applied to the Institution of Civil Engineers requesting the selection of an adjudicator.

The Institution of Civil Engineers selected Mr A. Wiseman, of 2 Oracle Road, Crawley, Sussex to be the Adjudicator.

The Background and Facts

EasyView invited tenders for The Newtown Franchise Area No 3 and on 6 January SuperConstruct Limited submitted their tender in the sum of £4,322,507.64.

Following submission of the tenders EasyView invited SuperConstruct to a meeting on 13 January. At the meeting Easyview informed SuperConstruct that the price had to be reduced by around 15% if the contract was to go ahead. The contract contained a bill of approximate quantities which were expressly stated to be for tender comparison purposes only.

Each item in the bill was discussed in turn and the rates for the trench work were all reduced by varying amounts.

It is about the values and method of measurement of Item 15 that has caused the dispute. Item 15 reads:

Backfill trench with imported material as directed;

The Unit of measurement is: M3 and the quoted rate is £15.

Although the tender documents did not expressly say so it was agreed at tender stage that this item should be priced on a place only basis and that the materials would be paid for in addition. A statement by our estimator is attached [Doc A] which records his conversation with Mr Jones, EasyView's surveyor who drafted the tender documents. There are three different possible backfilling materials permitted by the specification: selected backfill, free-draining material and lean mix concrete. It would be possible to supply the first two materials for the quoted rate but not the lean mix concrete. However that would leave insufficient money for the distribution of the materials, wastage and their removal of the displaced excavated materials.

At the meeting EasyView requested Superconstruct to agree to include the supply of the materials within the quoted rate. They said that it was very unlikely that there would be much imported materials as the majority of the material forming the existing pavements were imported granular sub-base material. SuperConstruct agreed that they would examine the possibility of including the supply of the material but did not do so.

On 15th January SuperConstruct resubmitted their bill of quantities with the new prices for the trenchwork but left the item for the backfilling unchanged because they had not agreed to any change.

The rate and the basis of the rate therefore remain as they were at the time of tender and SuperConstruct are entitled to be paid for the supply of the materials in addition to the rate for backfilling.

Even if that were not the case SuperConstruct would still be entitled to payment for the materials because, in spite of their statement to the contrary, EasyView entered into an agreement with the local authority to have all the trenches backfilled with free-draining material. This agreement was a sop by EasyView to the local authority who have vigorously opposed the digging up of their streets for additional cables.

We ask you to tell EasyView that the rate for backfilling does not include the supply of the materials

The second measurement issue is how the items for backfilling the trench and resurfacing the roads and footpaths should be measured. The problem arises because there is no method of measurement referred to in the Contract. In the appendix to the form of tender the item which usually defines the method of measurement used in the preparation of the bills of quantities the words "m3, lm or time as agreed" has been written in [Doc B].

These words indicated that the actual quantity of work carried out should be measured. EasyView are maintaining that we should only be paid the net quantities shown on the drawings. They base this view on the requirements of a number of methods of measurement including CESMM3 and SMM7, the civil engineering and building methods of measurement.

We do not see the merit of this argument as EasyView could, if that is what they wanted, have used one of these methods of measurement to define the bill of quantities. They chose not to do so. It is too late to try to rectify their mistake now.

Unfortunately, they do not show a dimension for the trench width on the drawings. This dimension is required, if their interpretation of the contract is right, for the quantities of the backfilling materials and the area of resurfacing of trenches. Two ducts each 80mm id are laid side by side. The Contract Administrator insists that the width of trench to be measured should be the net 170mm., which only allows for the thickness of the duct. This is clearly absurd. It would be impossible to compact the backfill material in accordance with the specification if there was no space for the backfill material to reach the haunches to the duct. It also provides no width for the collars for the joints.

If EasyView wish to base their arguments on the measurement of quantities on what is usual in the standard methods of measurement we would ask that the trench width should be similarly dealt with and Definition Rule D1 of Class L in the Civil Engineering Standard Method of Measurement states:

> *Where no dimension is given in the contract the minimum trench width shall be the distance between the inside faces of the outer duct walls plus 500mm where that distance does not exceed 1m....*

In this case therefore the minimum trench width should be 580mm. We ask you to so rule [Doc C].

There are two reasons why this difference is important. Firstly, the quantity of imported material has been radically increased (50 times). Secondly, the area of surfacing that we have had to replace has been increased because the local authority's inspector has insisted that additional areas of the existing footpaths, which are in very poor condition, should be replaced. They say that the existing surfaces although not perfect have been rendered unusable by our activities. We refute this entirely.

The footpaths are in the same condition as we found them but the new surfacing on the trenches does show up the bad condition of the rest.

The specification says that the work has to be to the satisfaction of the Contract Administrator and the local authority. We consider that we were required by the specification to execute the work in accordance with the instructions from the local authority's inspectors. EasyView say that they did not delegate any power of variation to the local authority inspector and therefore we should not be paid for the work that we were instructed to do.

However, they told their own inspectors that the surfacing was to be to the satisfaction of the local authority inspectors. The result was their own inspectors were never around when the surfacing works were being carried out. They never took any interest in these works and it would have been impossible to work if we had not followed what the local inspectors required. The Contract Administrator knew that we were being instructed to carry out additional resurfacing but chose to ignore it. We informed him at the time that we expected to be paid for the additional work.

We ask you to instruct that we should be paid for all the work that we have done including the work that the local authority inspector instructed.

Summary

We ask you to decide that:

(a) The item for backfilling does not include the supply of the materials which should be paid in addition.
(b) The quantities measured should be the actual quantity of the work done, and
(c) The minimum trench width should be 580mm.

I.B.Noah,
Director
SuperConstruct Limited

27th July

SuperConstruct Limited

Re: Adjudication with EasyView Ltd
Newtown Franchise Area - Contract No 4.

STATEMENT BY MR R.FIGGINS.

I was the estimator for the tender for this contract.

The tender documents were received in the office on 15 December, and they were given to me that day. I was engaged upon the submission of a tender which went in on the 17 December so I was unable to look at the documents in any detail until 18 December.

Since tenders had to be submitted on the 6th January and there was the Christmas break from 22nd December until 5th January this gave me very little time to obtain prices for materials and subcontracts.

Since there was a comparatively small amount of backfill required this seemed the least important matter for detailed research so I contacted Mr Jones, the surveyor dealing with the contract for EasyView, and suggested to him that we should price the item on a place only basis.

He said that other tenderers had suggested the same thing and that he would consult his director and get back to us. He phoned back the next day and told us that it had been agreed that for tender purposes the rate should be for place only.

I said that if we got the job we could agree to the materials being supplied on an invoice plus 7½% basis. I did not think there would be any problem since the quantities would be measured on a work done basis in the absence of any method of measurement.

Mr Jones said that he thought that might be possible.

Our tender was submitted on 6 January on that basis.

Signed Date: 22nd July

R.Figgins.
Estimator.

A.Wiseman
MA, FICE,FCIOB, FCIArb, MAE

Arbitrator and Adjudicator

2 Oracle Way
Crawley
Sussex

Tel: 01666 3241
Fax: 01666 3242
Email: adjudicator @ delphi.com

Mr A Grade,
EasyView Cable Co. Ltd.,
Satellite House,
Great Sight Street,
London, E43A 2IC.

Mr I.M.Noah,
Director,
SuperConstruct Limited
Unit 4,
Bene Volent Trading Estate,
Greenlands,
Surrey OR7 8SA.

29 July

Adjudication between SuperConstruct Limited and EasyView Cable Co Ltd.
Newtown Franchise Area - Contract No.4

I acknowledge receipt on 28 July of the adjudication Referral Notice dated 27 July served by SuperConstruct Limited.

In respect of the Referral Notice:

1. I confirm that I informed the Institution of Civil Engineers on 22 July that I was willing and able to act in this adjudication. I further confirm that I informed SuperConstruct on 25 July that I was willing to accept the appointment.

2. I enclose my terms for acting which you will note are based upon the Adjudicator's Contract published with the ICE Adjudication Procedure (1997).

3. Having carefully read the Referral Notice I now direct:
   (a) If EasyView wish to respond to the Referral Notice by way of defence they should do so on or before 6 August.
   (b) SuperConstruct may then reply to that statement on or before 13 August. Any comments in this reply must be confined to the points made in EasyView's statements and SuperConstruct should not raise any further argument of their own.
   (c) I shall then reach my decision during week commencing 15 August.
   (d) I do not need to visit the site and I do not intend to hold a meeting unless both parties request a meeting, in which case I may make the holding of the meeting conditional on the parties agreeing to extend the period within which I am required to reach my decision.
   (e) Any communication sent to me by either party must be simultaneously sent to the other party.
   (f) Any comments by either party upon these directions should

Page 2 of a letter from A.Wiseman dated 29 July

simultaneously be sent to the other party.

(f) Any comments by either party upon these directions should be sent to me within two days of receipt of this letter. In the absence of any comments these instruction will not be varied by me.

Yours faithfully,

A.Wiseman
Adjudicator

EasyView Cable Company Limited

Satellite House
Great Sight Street
London. E43A 2IC.

Tel: 0171 999 0000
Fax: 0171 999 0001

Mr A. Wiseman,
2 Oracle Road,
Crawley,
Sussex.

5 August

Dear Sir,

Adjudiction between SuperConstruct Limited and EasyView Cable Co Ltd
Newtown Franchise Area Contract No 3.

We refer to the Referral Notice sent to you by SuperConstruct dated 27 July and we wish to reply as follows:

1. EasyView deny the Claims made by SuperConstruct in their entirety.

2. The first seven paragraphs are admitted except that SuperContract were not told that they had to reduce their prices by 15%. This was merely expressed as the target that EasyView was looking for.

3. With regard to item 15 it is agreed that our Mr Jones asked for the rates be completed on a place-only basis. It was always the intention of EasyView the eventual rate in the contract would be on the basis of supply and that is the reason why there are no supply items in the Bills of Quantities for the backfilling materials. It was due to lack of time during the tendering period that led Mr Jones to agree to a place-only basis.

   It is agreed that at the meeting on the 13 January SuperConstruct were asked to quote for this item on an inclusive basis, and they agreed to look into it and make clear in their re-tender what they were requiring. It is correct that there are three possible materials for backfilling permitted by the specification.

   SuperContract submitted their re-tender on 15 January and they did not include any separate items for the supply of the backfilling materials which they would have done if they had required the materials to be paid in addition. It was clear to us that they had agreed to include the supply of the materials within the previously quoted rate.

   We attach a statement, signed by Mr Jones, who was our representative for the tender and post tender stages of this contract detailing his recollection of the various conversations and the comments made at the meeting on 13 January.

We would also comment that it is the custom in the industry as exemplified by all standard methods of measurement that the prices for the items include all the activities and other resources (including materials) required to complete the components of work described. Therefore without further explanation the backfilling item as described in the Bill of Quantities would include the supply of the necessary materials unless supply was expressly excluded, which in this case it was not.

4. With regard to the manner in which the quantities should be calculated it is so obvious that it goes without saying that the contractor should only be paid for the work which he is required to do. This is defined by the drawings and the specification. If the exact requirement is not shown then it is the custom in the industry and defined by the industry's main Standard Methods of Measurements, SSM7 for building works and CESMM3 for civil engineering works, that the quantities are computed net.

The reason why no standard method of measurement was used for this contract was that this is specialist work which does not lend itself to the item descriptions in any of the standard methods.

Whilst it is agreed that EasyView did agree with the local authority that the trenches should be backfilled with free-draining material, that is not a matter for SuperConstruct. They tendered for the work and just because the quantities increase does not mean they are entitled to a different rate. The contract is an admeasurement contract and provides for variations, when they are similar to work for which there are rates in the Bills of Quantities the varied work should be measured at such rates.

5. With regard to the trench width, again this is a simple matter of definition. All that SuperConstruct were required to do was to dig a trench wide enough for the ducts to be placed in. Any additional width is clearly for their purposes and not those of the client. If they are unable to compact the bedding material with the equipment they wish to use without digging wider than the minimum required to place the ducts that again is their business. The minimum trench width for the two side-by-side ducts must therefore be the minimum width necessary to hold the ducts. The two ducts were each 80mm id. ducts and to this must be added the duct wall thicknesses, say 2mm, giving a total trench width of 160mm + 4x2mm = 168, say 170mm.

The amount demanded by SuperConstruct is outrageous and indicates why CESMM3 was unsuitable for this type of work. The suggestion that it should be used for the basis of the trench width is specifically denied.

6. With regard to the surfacing, again we refute that SuperConstruct are entitled to any additional payment over the net trench width. It is admitted that the specification says that the work is to be to the satisfaction of the local authority, but that means the work that SuperConstruct are required to do under the Contract. Not work that they choose to do. The Contract is quite specific.

By Clause D4:

"The Contractor shall not vary the Works in any manner whatsoever without an

instruction from the Contract Administrator. All such instructions shall be in writing unless the Contract Administrator deems it necessary to give such an instruction orally, in which case the Contract Administrator will confirm such oral instruction within 24 hours. "

By Clause D5:

"The Contract Administrator may delegate his power to issue variation instructions. Any such delegation shall be in writing and shall not take effect until such time as the Contractor is notified in writing. Such delegation may be revoked at any time and shall take effect from when the contractor is informed of such delegation whether that be in writing or not."

No delegated authority was ever given to the local authority inspectors. No written notification could therefore have been sent to SuperConstruct, nor was it ever claimed by SuperConsult to have been received.

If SuperConstruct did any additional work outside the net trench width they did it either because they excavated additional width or they damaged the existing surfacing. In either case that is nothing to do with EasyView.

It is agreed that SuperConstruct informed the Contract Administrator about the additional surfacing, but it is denied that he ignored this. On the contrary, he replied that SuperContract would be paid if the work had been carried out in accordance with the requirements of the Contract. It was not, so SuperConstruct are not entitled to any additional payment.

7. Summary:

We request that you find:

(a) The backfilling items include for the supply of the materials.
(b) The quantities should all be measured net from the drawings and where the drawings do not define the extent width exactly then the quantity measured should be the minimum necessary to fulfill the contractual obligations of the contractor. In the case of the surfacing and backfill it is the "minimum trench width".
(c) The minimum trench width for the 2-duct runs should be 170mm.

If you require any further information please contact us.

Yours faithfully,

B.Grade,
Director,
EasyView Cable Co.

cc SuperConstruct Ltd
Painters
S. Tight

24 East Street,
Dundee DN24 8JT.

☎ 0145 1452

Adjudication between SuperConstruct Limited and EasyView Cable Co Ltd
Newtown Franchise Area Contract No 3.

STATEMENT BY Mr B.JONES.

I have read the Referral Notice served by SuperConstruct on EasyView on 26 July and I have been asked to give my recollection of the events described in it with which I was involved.

I am a self-employed freelance quantity surveyor. I have been an Associate of the Royal Institution of Chartered Surveyors since 1956. I have been involved most of my working life in the production of contract documents and the measurement of construction work.

In November I was asked by EasyView to prepare the tender documents for the Newtown No.3 Area. This I did and sent out the documents to a selected list of contractors on 13 December, one of which was SuperConstruct.

During the tender period, which was only just over three weeks and included the Christmas break, I was contacted by a number of contractors concerning Item 15 in the Bill of Quantities concerning backfill. They asked me how they should estimate the quantities for the various types of backfill. Because the layout design of the system was at that stage in a fairly embryonic stage I was unable to give them any advice, and after a number of discussions felt that the only sensible way to proceed was for the tenderers to price for place only at tender stage for comparison purposes since the actual supply would be very much the same whichever contractor was awarded the contract.

I informed the contractors that they should therefore quote on a supply only basis but that they would be required to negotiate a supply and place rate if their tender was otherwise acceptable.

I opened the tenders on 6 January, and SuperConstruct were not the lowest. I contacted their Mr Holroyd, who was responsible for their tenders, and asked him to attend a meeting on 13 January to discuss their tender. This they agreed to do.

At the meeting which was attended by myself and Mr Holroyd I said that the client was looking for a saving of 15% on their figures. Mr Holroyd was clearly anxious to get the job and agreed to reduce his rates for the trenchwork and surfacing by around 15%. without any persuasion.

When it came to the backfilling item he was worried about the proportions of the type of backfill material, because he said he could not obtain the lean mix concrete for the quoted price. I told him that there would be little lean mix backfill, which turned out to be true. In fact there was none. He said that in that case he thought he could stand by the rate for backfill.

I then asked him to resubmit his Bill of Quantities with the new rates and making any other alterations that he considered necessary.

I received the revised Bill of Quantities the following day. The rates had been reduced by approximately 15% and the item for backfill had not been changed. In accordance with my instruction that he should make any alteration necessary it was clear that he had accepted that the item for backfilling should include the supply of the materials.

We did not discuss the way the items should be measured and if Mr Holroyd had any queries he should have raised them at the meeting since I asked him specifically if there was anything that was not clear because I had had to produce the documents in a very short time. He said he had no queries. It is too late to raise them now.

I produced the formal contract documents which were signed on 13 February.

I then ceased to be involved with the project since I had completed the work that I had been engaged to do.

This is a true statement to the best of my recollection and knowledge.

B. Jones.
Chartered Surveyor

2 August

SuperConstruct Limited

Telephone: 01245 100100
Facsimile: 01245 100101

Unit 4
Bene Volent Trading Estate
Greenlands
Surrey OR7 8SA

Mr A. Wiseman,
2 Oracle Road,
Crawley,
Sussex.

12 August

Dear Sir,

Adjudication between SuperConstruct Limited and EasyView Cable Co Ltd
Newtown Franchise Area Contract No 3.

We refer to the Reply to the Referral Notice, dated 5 August, sent to you by EasyView.

In accordance with your instruction 3(b) in your letter dated 29 July, we reply as follows:

1. Re Item 2.
Our interpretation was that this was a bit more formal and we enclose a statement by our Mr Holroyd who was at the meeting.

2. Re Item 3. 3rd Paragraph.
We deny that we were asked to re-tender on an inclusive basis. We were asked if we would supply the materials for the quoted price and we said we would look into it. No statement was made that we should make clear when we re-tendered the basis of this item. We were only asked to reprice the items in the Bill of Quantities and a copy of it was provided for us to price. Again Mr Holroyd confirms this in his statement.

   We deny that we were asked to include prices for the supply of the materials. It was never suggested that we should add items to the Bill of Quantities.

   5th Paragraph.
We deny that custom in the industry is relevant. It was agreed between the parties what the basis of the tender was and there is no evidence that this was ever changed.

3. Re Item 4.
We deny that either CESMM3 or SMM7 are relevant. In our experience if a standard method of measurement is not specified the draftsman of the Bill normally writes his own and defines how the quantities are to be computed.

   In the absence of any statement in the Bill it can be said that the Bill is ambiguous and should be interpreted strictly contra proferentem, that is against the drafter.

   3rd Paragraph.
Whilst the contract may not entitle us to a changed rate, nevertheless we were informed by Mr Jones, on behalf of EasyView, that there would be little imported material used and we tendered on this basis. The agreement with the local authority renders this a misrepresentation within the

meaning of the Misrepresentations Act 1967. Although the remedy for an innocent misrepresentation is recission, a court has the power to award damages and so would you.

4. Re Item 5.
EasyView intentionally do not say that the bedding materials are required to be compacted all around both ducts. We must therefore be able to compact it. This is not possible if it is impossible for the bedding material to be able to be provided to fill any voids that may be formed on the underside of the ducts by the compacting action. The aggregate for the bedding is 25mm Type 2 and sufficient width must be provided to allow access. Individual bits of aggregate with dimensions larger that its nominal size will be able to pass the screen and therefore additional width should be provided to accommodate this. We would suggest that a minimum figure of 35mm should be used.

5. Re Item 6.
The implication from the last paragraph is that SuperConstruct did not carry out the work in accordance with the contract. We did precisely what we were told to do by the local authority inspector and our work was to be carried out "to the satisfaction of the local authority". EasyView on all occasions went out of their way to agree to anything that the local authority required for their own political reasons. They totally abdicated all responsibility for sorting out the excess demands of the local authority inspectors and unless we agreed to do what the inspectors asked us to do we were unable to get the work passed by them. We were expressly told to do as the inspectors required.

6. Re Item 7.
We ask that you find as requested in our Referral Notice.

Yours faithfully,

I.B.Noah,
Director
SuperConstruct Limited

cc. EasyView Cable Company Limited
J. Holroyd

SuperConstruct Limited

Adjudication between SuperConstruct Limited and
EasyView Cable Co Ltd.
Newtown Franchise Area - Contract No. 3

STATEMENT BY Mr J.HOLROYD

I am the director of SuperConstuct in charge of Estimating. I oversaw the production and submission of our tender for this contract.

I have read the Reply to our Referral Notice provided by EasyView Cable Company Limited and make the following observations:

1. Item 2.
   My recollection is that we were asked to reduce our price by 15% if we were to get the job. They said that we were the preferred contractor and would be awarded the contract with the 15% which would still not make us the lowest. I know that the whole emphasis of the discussion on the rates was could we take 15% off each one.

2. Item 3
   Again my recollection is that we were asked to re-tender and we were given a blank copy of the Bill of Quantities to return. We were asked to quote on an inclusive basis and I said I would look into it because we had not received any quotes for the free-draining or selected fill materials. I knew approximately what the lean mix would cost, about three times as much as our quoted rate.

   I gave no undertaking that we could include the materials. We were certainly not asked to "make clear in our re-tender what we were requiring." It was never suggested that we write in extra items for the supply of the materials as this had been discussed as being likely to be cost plus 7½% if we could not include the supply in our quoted rate.

   We were merely asked to submit the Bill as printed. This we did.

   We left it that we would inform them if we were able to include the materials supply in the rate. We were not so we did not contact them.

   Since we made no change to the basis of our original quote it was not necessary to make any alteration to the descriptions, and neither were we asked to.

   That to the best of my recollection is what happened at the meeting on 13 January.

J.Holroyd. 10 August

A.Wiseman
MA, FICE,FCIOB, FCIArb, MAE

Arbitrator and Adjudicator

2 Oracle Way
Crawley
Sussex

Tel: 01666 3241
Fax: 01666 3242
Email: adjudicator @ delphi.com

Mr A Grade,
EasyView Cable Co. Ltd.,
Satellite House,
Great Sight Street,
London, E43A 2IC.

19 August

Mr Noah,
Director,
SuperConstruct Limited
Unit 4,
Bene Volent Trading Estate,
Greenlands,
Surrey OR 8SA.

Adjudication between SuperConstruct Limited and
EasyView Cable Co Ltd.
Newtown Franchise Area - Contract No.4.

I enclose my decision in this Adjudication.

I also enclose with each party's copy of this letter an invoice in respect of my fees and expenses which I have determined to be £1,895.2 including £282.27 VAT. I have also determined in my decision that these fees and expenses should be borne equally by the parties.

Yours faithfully,

A.Wiseman
Adjudicator

An

Adjudication

between

SuperConstruct Limited

and

EasyView Cable Company Ltd.

Newtown Franchise Area - Contract No. 3.

DECISION

A.Wiseman
Adjudicator
2 Oracle Way,
Crawley,
Sussex.

19 August

Adjudication between SuperConstruct Limited [SuperConstruct] and EasyView Cable Co Ltd [EasyView].

Newtown Franchise Area - Contract No. 3.

ADJUDICATOR'S DECISION

Introduction

1. EasyView Cable Co Ltd provides cable access to domestic and business premises for the purposes of communications and television. They won in competition the franchise to provide such services in Newtown. SuperConstruct Ltd is a firm of contractors experienced in laying cable ducts.

2. The parties entered into a Contract, known as the Newtown Franchise Area - Contract No 3, for the laying of approximately 2Km of cable ducts on 13 February in the sum of £3,823,207.34.

3. Work commenced on 23 February and within a very short period it became apparent that there were fundamental differences between the parties as to the interpretation of the contract between them.

4. The parties held a series of meetings but were unable to resolve their differences and SuperConstruct on 20 July gave Notice of Adjudication under the Housing Grants Construction and Regeneration Act 1996. There is no adjudication provision in the contract.

5. SuperConstruct also on 20 July asked the Institution of Civil Engineers, which is a body which holds itself out publicly as a body which will select an adjudicator when requested to do so by a referring party, to select an adjudicator.

6. On 24 July the Institution of Civil Engineers rang me and asked if I would be willing and able to act in this adjudication and I confirmed that I would. On 25 July the Institution of Civil Engineers informed the parties that they had selected me, Abel Wiseman, of 2 Oracle Way, Crawley, Sussex, as adjudicator.

7. On 25 July SuperConstruct rang me and asked if I was able and willing to act as adjudicator and again I confirmed that I was. On 27 July SuperConstruct sent the Referral Notice to me, which I received on 28 July.

8. Having carefully read the Referral Notice and the Notice of Adjudication I confirmed my acceptance of the appointment in writing to the parties and issued certain directions for the conduct of the adjudication.

9. EasyView submitted their statement and views on the dispute to me on 8 August and then SuperConstruct submitted comments on that statement on 15 August.

10. I now make my decision on the basis of these communications.

The Issues

10. I have been asked to decide three issues that:

(a) EasyView allege the item for backfilling in the bill of quantities includes the supply of the necessary materials and SuperConstruct allege that it does not and the materials should be paid for in addition.
(b) SuperConstruct claim that they should be measured for the quantities of work that they actually executed whereas Easy View allege that the quantities should be the minimum necessary to carry out what they require to be done.
(c) SuperConstruct claim that the "minimum trench width" for the ducts should be 580mm whereas EasyView consider that it should be 170mm.

11. <u>Issue (a)</u>
The Bill of Quantities incorporated into the Contract contained an item, Item 15, with the description "Backfill trench with imported material as directed". A dispute arose as to whether or not this item included the supply of the necessary materials.

The parties are agreed that at SuperConstruct's rate this item in their tender submitted on 6 January did not include the supply of the materials.

On 13 January EasyView called a meeting with the purpose of reducing the rates tendered by SuperConstruct. These were in fact reduced by approximately 15 %. When it came to item 15 the parties' recollection of what occurred is diametrically opposed. EasyView say that SuperConstruct agreed to supply the materials for the quoted rate and SuperConstruct say they did not.

At the end of the meeting EasyView say that they asked SuperConstruct to resubmit the Bill of Quantities with the new rates and to include any changes that they wished to have made in the Bills of Quantities. Again SuperConstruct say that no such instruction was ever given to them.

On 15 January SuperConstruct resubmitted their Bill of Quantities but Item 15 remained unaltered. SuperConstruct argue that this indicated that SuperConstruct had agreed to include the supply of the materials and SuperConstruct say that it merely showed that the basis of their tender remained unaltered.

Both parties provided a number of additional arguments in support of their positions but it seems to me that it is for the party who is claiming that there has been a change to the agreed basis of the rate to prove on the basis of the balance of probability that their position is the more likely. The burden therefore in my opinion rests with EasyView to show that the basis was changed and I do not consider that they have satisfied this requirement.

I therefore decide that the rate for Item 15 does not include for the supply of the materials.

12. Issue (b)

The Bill of Quantities was not drafted in accordance with any standard method of measurement and a difference of opinion has arisen between the parties as to whether the quantities should be computed net or be computed by measuring the actual quantity of work done. This affects two items in particular, namely, the area of surfacing replaced and the quantity of backfill.

In my opinion, it is easiest to consider this issue in two stages: first the general principle and then the specific components of work.

It cannot be right that the contractor should be paid for all the work that he does whether or not he was ordered to do it. He should only be measured for the work that is necessary in order to fulfill his contractual obligations. In the absence of a method of measurement that does not necessarily mean net from the drawings if additional work is necessary to meet the requirements of the specification.

With regard to the surfacing measurement SuperConstruct claim that they have been ordered to carry out a large amount of excess surfacing at the instigation of the local authority inspectors, who say that the existing surfacing has been damaged as a result of SuperConstruct's operations. EasyView on the other hand say that it is no business of theirs and they will only pay for what is expressly required by the contract. Further they have given no authority to the local authority inspectors to issue variations orders.

I reject both arguments. It seems to me that the starting point for payment is that the measurement should be based on the minimum trench width as decided in Item (c) below. However, the work has to be acceptable to the local authority inspector. It is clearly impractical for SuperConstruct to have to send for the Contract Administrator every time that there is an argument about whether excess surfacing is required as a result of damage or for betterment. If EasyView are not prepared to have their own inspector present to decide such matters it seems to me that it would have to be implied that they had delegated such power to the local authority's inspectors.

However, it must be for SuperConstruct to show that they have not damaged the existing surfacing by their operations. If they cannot do that then the cost of the additional surfacing must be executed at their expense.

If they can show that the surfacing ordered to be replaced was in exactly the same condition as they found it then the cost of the extra resurfacing to obtain the local authority's inspector's satisfaction must be paid for by EasyView.

13. Issue (c)

On the basis that the correct principle is that the net trench width should be the minimum width necessary to enable the requirements of the specification to be met, I reject the arguments of both parties.

In my view the width required must be made up of the maximum diameter of the ducts (that is at their collars) plus sufficient width to ensure that the bedding material is compacted in accordance with the specification. In this case the specification required 25mm Type 2 aggregate. The minimum width must therefore allow for the maximum size of bedding material 25mm.

I am therefore of the opinion that the width of trench to be measured should be the sum of the external diameter of the ducts (say 84mm) plus the thickness of the collars (say 4mm) plus the space around the ducts necessary for compaction (say 25mm. For a double 80mm id duct this gives a minimum trench width of 2 x 84mm plus 2x 4mm plus 3 x 25mm which equals 251mm, say 250mm. This width should normally be used for calculating the quantities of surfacing and backfilling.

The Decision

14. I therefore decide that
    (a) The backfill rate item 15 of the Bill of Quantities did not include the supply of the necessary materials.
    (b) The quantities measured should be computed on the basis of the minimum reasonably necessary to fulfill the requirements of the specification.
    (c) The minimum trench width for the double 80mm id duct should be 250mm.

15. Because these decisions are fairly evenly divided between the arguments of the parties I apportion my fees and expenses equally between the parties.

A.Wiseman
<u>Adjudicator</u>

19 August

ICE Conditions of Contract for Minor Works

APPENDIX TO THE CONDITIONS OF CONTRACT

(to be prepared before tenders are invited and to be included with the documents supplied to prospective tenderers)

1. Short description of the work to be carried out under the Contract
   Approx 200km of double 80mm id cable duct and house connections
   ....

2. The payment to be made under Article 2 of the Agreement in accordance with Clause 7 will be ascertained on the following basis. (The alternatives not being used are to be deleted. Two or more bases for payment may be used on one Contract.)

   (a) ~~Lump sum~~

   (b) Measure and value using a priced Bill of Quantities

   (c) ~~Valuation based on a Schedule of Rates (with an indication in the Schedule of the approximate quantities of major items)~~

   (d) ~~Valuation based on a Daywork Schedule~~

   (e) ~~Cost plus fee (the cost is to be specifically defined in the Contract and will exclude off-site overheads and profit)~~

3. Where a Bill of Quantities or a Schedule of Rates is provided the method of measurement used is
   m^3, Lm or time as agreed

4. Name of Engineer (Clause 2.1)
   Mr S.O. Tight

5. Starting date (if known) (Clause 4.1)
   22nd February

6. Period for completion (Clause 4.2)
   6 months

7. Period for completion of parts of the Works if applicable and details of the work to be carried out within each such part (Clause 4.2)

	Details of work	Period for completion
Part A	....	....
Part B	....	....
Part C	....	....

8. Liquidated damages (Clause 4.6)
......£10,000 per week......

9. Limit of liquidated damages (Clause 4.6)
......£ 500,000......

10. Defects Correction Period (Clause 5.1)
......12 Months......

11. Rate of retention (Clause 7.3)
......10%......

12. Limit of retention (Clause 7.3)
......7½%......

13. Minimum amount of interim certificate (Clause 7.3)
......£50,000......

14. Bank whose base lending rate is to be used (Clause 7.8)
......Barclays Bank Limited......

15. Insurance of the Works (Clause 10.1)
Required/~~Not required~~

16. Minimum amount of third party insurance (persons and property) (Clause 10.6)
......£2m......
Any one accident/Number of accidents unlimited

17. Name of the Planning Supervisor (Clause 13(1)(b))
......Mr S.O. Tight......
Address ......(See Engineer)......

18. Name of the Principal Contractor (Clause 13(1)(b))
......The Contractor.......
Address ......

CLASS I

MEASUREMENT RULES	DEFINITION RULES	COVERAGE RULES	ADDITIONAL DESCRIPTION RULES
M1 The Commencing Surface adopted in the preparation of the Bill of Quantities shall be adopted for the measurement of the completed work. **M2** Backfilling of trenches shall not be measured except as set out in class K for filling of French and rubble drains and in class L for backfilling with material other than that excavated from the trenches. **M3** Lengths of pipes shall be measured along their centre lines. Lengths of pipes *in trenches* shall include lengths occupied by fittings and valves and exclude lengths occupied by pipes and fittings comprising backdrops to manholes. Lengths of pipes *not in trenches* shall exclude lengths occupied by fittings and valves. **M4** Additional items shall be given in classes K and L for work in connection with pipes *not in trenches* other than the provision, laying and jointing of pipes. **M5** Lengths of pipes entering manholes and other chambers shall be measured to the inside surfaces of the chambers except that pipes and fittings comprising backdrops to manholes shall be included in items for manholes measured in class K.	**D1** Pipes *not in trenches* shall include pipes suspended or supported above the ground or other surface, pipes in headings, tunnels or shafts, pipes installed by thrust boring and pipe jacking and pipes laid within volumes measured separately for excavation. **D2** Pipes *not in trenches* shall be classed as such only where pipes are expressly required not to be laid in trenches. **D3** Depths used for classification of pipes *in trenches* shall be measured from the Commencing Surface to the inverts of the pipes.	**C1** Items for pipes shall be deemed to include the supply of all materials by the Contractor unless otherwise stated. Items shall be deemed to include pipe cutting. **C2** Items for pipes *in trenches* shall be deemed to include excavation, preparation of surfaces, disposal of excavated material, upholding sides of excavation, backfilling and removal of dead services except to the extent that such work is included in classes J, K and L.	**A1** The location or type of pipework in each item or group of items shall be stated in item descriptions so that the pipe runs included can be identified by reference to the Drawings. **A2** The materials, joint types, nominal bores and lining requirements of pipes shall be stated in item descriptions and reference given to applicable British Standard specifications and specified qualities. **A3** Item descriptions for pipes *not in trenches* shall distinguish between the different categories of pipes listed in rule D1. **A4** The Commencing Surface shall be identified in the description of each item for work involving excavation for which the Commencing Surface is not also the Original Surface. **A5** Where more than one pipe is expressly required to be laid in one trench the item descriptions for each pipe shall so state and also identify the pipe run. Where pipes are laid in French or rubble drains item descriptions shall so state. **A6** Trench depths exceeding 4 m shall be stated in item descriptions to the next higher multiple of 0·5 m. **A7** Item descriptions shall identify separately excavation which is expressly required to be carried out by hand.

Appendix E

Construction Industry Council's Guidance Notes for Adjudicators

These notes have been prepared by the Construction Umbrella Bodies Adjudication Task Force consisting of members from:

- The British Property Federation (BPF)
- The Construction Federation (CC)
- The Construction Industry Council (CIC)
- The National Specialist Contractors Council (NSCC)
- The Specialist Engineering Contractors' Group (SECG)
- The Construction Sector Unit of the Department of Trade and Industry (DTI)

The author was a member of the Task Force. Copies can be obtained from the Construction Industry Council, 26 Store Street, London WC1E 7BT.

Adjudication under the Scheme — Guidance to Adjudicators

Contents

Introduction

This guidance to adjudicators arises out of the Construction Industry Board's review[1] of the first two years' operation of the Scheme[2] under the Construction Act.[3] It covers a number of specific issues raised in that report, and is not comprehensive.

Circumstances alter cases, and the guidance should be treated as suggestions to adjudicators rather than rules. It is couched in lay terms and is not intended as a legal interpretation of the provisions of the Act or the Scheme, but as helpful guidance. *The guidance is advisory not binding.*

The guidance primarily refers to adjudication conducted under the Scheme, but much of it will be applicable when a contractual adjudication procedure is used.

1. Natural justice (procedural fairness)

The issue

An adjudicator must conduct the proceedings in accordance with the requirements of natural justice or procedural fairness. In a small number of cases, the courts have not enforced the adjudicator's decision on the grounds that the adjudicator did not act fairly, and it is evident that adjudicators do not always know exactly what it means to act in accordance with natural justice.

The law

Natural justice is not a defined term. As one judge put it: "Natural justice requires that the procedure before any tribunal which is

1 *Review of the Scheme for Construction Contracts*: A CIB Report to the Construction Minister, December 2000: available (together with the DETR Response) on www.ciboard.org.uk

2 The Scheme for Construction Contracts (England and Wales) Regulations 1998 (Statutory Instrument 1998 No. 649).

3 Part II of the Housing Grants, Construction and Regeneration Act 1996 which came into force on 1 May 1998.

acting judicially shall be fair in all the circumstances, and I would be very sorry to see this fundamental principle degenerated into hard-and-fast rules."[4]

There are two main aspects to the need for procedural fairness: "no bias" and "fair hearing".

No bias: the adjudicator must be impartial and act independently. For there to be a breach of natural justice it is not necessary for there to be actual bias — apparent bias is sufficient. The test is whether there is a real *possibility*, not probability, that the adjudicator is biased based on how the reasonable observer would interpret the situation. Bias may occur in a number of ways; for example, if the adjudicator

- has, or appears to have, a personal relationship with one of the parties
- has, or appears to have, an interest in the outcome of the adjudication (a conflict of interests)
- conducts the adjudication in a manner which favours, or seems to favour, one party
- acts in a manner which is seen, or might be seen, as supporting one party to the detriment of the other.

Fair hearing: this means ensuring that each party:

- has a reasonable opportunity of presenting its case;
- knows what the case is against it;
- is in possession of all the evidence and information that is adduced against it or obtained by the adjudicator.

The words do not imply that an oral hearing or meeting is necessary. These requirements have to be measured in the context of the time within which the adjudicator has to reach his decision, and the fact that the decision is provisional, pending final resolution of the dispute by arbitration, litigation or agreement. As one judge put it:

4 *Wiseman* v. *Borneman* [1971] AC 297 HL, per Lord Reid at 308B.

"It is accepted that the adjudicator has to conduct the proceedings in accordance with the rules of natural justice or as fairly as the limitations imposed by Parliament permit."[5]

Guidance to adjudicators

1. If you are aware of any connection, however remote, that you have or have had with either party, or of any matter which either party might see as being a conflict of interests, notify the parties and consider refusing the appointment. If anything comes to light after you have been appointed, ask the parties if they wish you to continue; if necessary, resign.[6]
2. Ensure that all your actions are, and are seen to be, fair. Before taking any action, (for example, declining to accept late information)[7] ask yourself: "Am I acting fairly? Does it appear that I am acting unfairly?"
3. Although the Scheme does not make express provision for the submission of a response to the referral notice, give the responding party a reasonable opportunity to respond, and consider any response that is served.[8] You may also instruct that a response is served, and specify the timetable for this to be done.[9]
4. Use great care if you communicate with one party (or its representatives, experts or witnesses) in the absence of the other, either in a meeting or over the telephone. If contact with one party alone is necessary or unavoidable, keep a detailed record of what is said and send it to both parties as soon as practicable. If you receive unsolicited telephone calls, consider requiring that the information is conveyed in a form which can be

5 *Glencot Development & Design Company Ltd* v. *Ben Barrett & Son (Contractors) Ltd* [2001] BLR 207, TCC, Judge Humphrey LLoyd; see also *Discain Project Services Ltd* v. *Opecprime Development Ltd* [2001] BLR 285, TCC, Judge Bowsher.

6 The Scheme, paragraph 9(1).

7 The Scheme, paragraph 15.

8 The Scheme, paragraph 17.

9 The Scheme, paragraph 13(a).

sent to the other party and yourself, for example by fax or email. Ensure that both parties know what is happening at all times.

5. In the case of necessary telephone calls, consider using telephone conferencing.
6. Consider obtaining the views of the parties before making directions other than those you consider routine.
7. Notify as early as possible final deadlines for the supply of information and dates for meetings.
8. Although your role is investigatory and inquisitorial, when undertaking investigations, ask yourself whether you are attempting to make or supplement one party's case; do not argue the case for one of the parties, either before or when giving your decision, or in setting out your reasons.
9. Consider carefully before indicating disapproval of the way in which one party conducts his case.
10. Guard against seeming to have made up your mind as to the merits of the case before having heard or seen all the evidence.
11. Ensure that any information on which you intend to rely in reaching your decision is known to both parties, so that they both have an opportunity of responding.

2. Challenges to jurisdiction

The issue

Almost the only way that a party can stop an adjudication proceeding is by alleging that the adjudicator lacks the jurisdiction — that is, the authority — to deal with the issues raised. If an adjudicator proceeds without the necessary jurisdiction his decision will not be enforceable. On the other hand if he fails to proceed when he does have jurisdiction, that is unjust to the referring party.

Examples of grounds on which jurisdictional challenges may be made are:

- the contract is not a "construction contract" within the Act;[10]

10 The Act, sections 105 to 107.

- the relevant activities are not "construction operations" within the Act;[11]
- the adjudicator was not properly appointed, for example because he is not the adjudicator named in the contract, the wrong ANB made the appointment, the appointment was too late, or the adjudicator has a conflict of interests;
- it is asserted that there is no dispute.

The law

Unless the parties otherwise agree, an adjudicator does not have the power to make a final decision as to whether he has jurisdiction to act as an adjudicator under the Scheme; only the courts can do that. However, as one judge put it: "It is clearly prudent, indeed desirable, for an adjudicator faced with a jurisdictional challenge which is not a frivolous one to investigate his own jurisdiction and to reach his own non-binding conclusion as to that challenge. An adjudicator would find it hard to comply with the statutory duty of impartiality if he or she ignored such a challenge."[12]

Guidance to adjudicators

1. If you are faced with a jurisdictional challenge, you should investigate, seek the views of the parties and reach your own conclusion on the merits of the challenge. If you fail to do so, it may seem that you are not impartial.
2. If you conclude that you do have jurisdiction, you should tell the parties immediately and continue with the adjudication.
3. If you conclude that you do not have jurisdiction, you should tell the parties immediately and give notice in writing of your intention to resign.[13]

---

11 The Act, section 105.

12 *Christiani & Nielsen Ltd* v. *The Lowry Centre Development Company Ltd*, unreported, 29 June 2000, per Judge Thornton.

13 The Scheme, paragraph 9(1).

4. If you are unsure whether you have jurisdiction, nevertheless make a judgment as to whether to proceed or resign.
5. If you proceed, consider obtaining confirmation from the referring party that he wishes you to continue.

3. Intimidatory tactics

The issue

There is growing evidence that some adjudicators are experiencing intimidatory or "bullying" tactics from the parties or their representatives during the course of an adjudication.

For example, undue pressure may be put on the adjudicator to adopt a course of action desired by one of the parties. This may be by means of aggressive threats or other actions designed to reduce the control that the adjudicator has over the process. It may involve:

- making spurious challenges to the adjudicator's jurisdiction;
- causing delay with the intention of obtaining an extension of time;
- deliberately confusing the adjudicator through the use of technical or esoteric legal arguments;
- threatening to take no further part in the adjudication or to take legal action against the adjudicator, or to report him to his professional institution.

The law

Paragraph 13 of the Scheme lists some of the steps that an adjudicator can take in order to determine the dispute, and paragraph 14 requires the parties to comply with any request or direction of the adjudicator.

Guidance to adjudicators

1. Recognize such tactics early and counter them firmly but fairly. Remember that you are in control of the procedure.

2. Where necessary, require that all arguments are put in writing with a copy sent to the other party.
3. Follow the advice regarding telephone calls in section 2, that a detailed record be kept. Consider refusing to discuss any matter over the telephone.
4. Respond politely but firmly if a party is behaving in an intimidatory manner; it may be appropriate, where arguments are being put forward aggressively or at undue length, to limit the length of submissions that may be made.
5. Always remember that while issues raised by the parties should be taken into account, they or their representatives should not deflect you from the duties placed upon you.
6. If one party goes to court, for example to challenge jurisdiction, continue the adjudication unless both parties agree otherwise or the court so directs.
7. Remember that adjudication is a rapid procedure leading to a provisional decision and that, if need be, the complexities of the underlying dispute can be resolved in arbitration or litigation.
8. Do not lose your temper!

4. Unmanageable documentation

The issue

Sometimes one party submits unmanageable and disproportionate quantities of documentation to the adjudicator, making it difficult for the other party to respond and for the adjudicator to reach his decision within the time limit.

Some adjudicators express concern that if they seek to limit the amount of paperwork that they receive or consider, they will be vulnerable to challenge either on the grounds of natural justice or because it would be a breach of paragraph 17 of the Scheme (which requires the adjudicator to consider any relevant information). They (wrongly) believe that they must take into account *all* information submitted to them in the course of the adjudication.

The law

As explained, natural justice requires (amongst other things) that the adjudicator must give each party a reasonable opportunity to present its case and act fairly between the parties. However, "fairness" must be set within the context of adjudication as a fast and interim procedure.

Paragraph 17 of the Scheme requires the adjudicator to consider "any *relevant* information submitted to him". Paragraph 13 gives the adjudicator power to limit the length of written documents submitted to him.

Guidance to adjudicators

1. The responsibility for judging what is or is not "relevant information" lies with you but in principle it is information that is evidential of the issues or events that a party has to prove in order to further its case.
2. The duty in paragraph 17 also needs to be considered in the light of the extensive powers given to you by paragraph 13. In particular consider giving directions as to the timetable, any deadlines, or limits as to the length or quantity of written documents.
3. Consider seeking consent to an extension of time in order to give you a proper opportunity to consider the papers and the other party's response.[14]
4. Remember that you are in control of the procedure. Consider limiting *at the outset* the amount of material to be submitted, taking into account the nature and value of the dispute. To prevent documents being "drip-fed", you can stipulate that all submissions and all documents must be received by a certain date.
5. Consider requiring a party to provide a concise statement of its case, cross referenced to a bundle of back-up documentation, and a chronology (although an extension of time may also be needed).

14 The Scheme, paragraph 19(1).

6. Consider whether the documentation properly relates to the dispute being adjudicated or to another issue (which may or may not be a dispute between the parties).
7. Consider your powers under paragraph 15 of the Scheme where a party declines to comply with a request without showing sufficient cause.

5. Reasons for the Decision

The issue

The adjudicator is obliged to give reasons for his decision if requested by one of the parties. The possibility of receiving a request at a late stage during the adjudication, or even after the decision has been given, has caused concern amongst adjudicators.

The law

Paragraph 22 of the Scheme provides: "If requested by one of the parties to the dispute, the adjudicator shall provide reasons for his decision". The adjudicator also has a wide discretion, under paragraph 13(g) and (h) to "give directions as to the timetable for the adjudication, any deadlines..." and "issue other directions relating to the conduct of the adjudication".

Guidance to adjudicators

Consider setting a date at the outset by which any request to give reasons must be made.

6. Accidental errors or omissions

The issue

Uncertainty has arisen regarding the extent to which an adjudicator may correct accidental errors or omissions (errors) in his decision once it has been delivered to the parties.

The law

Once the adjudicator has delivered his decision, his jurisdiction over the dispute is ended. However, where there is an error on the face of the decision, it seems that the adjudicator retains a power to make corrections. As one judge put it:

"... in the absence of a specific agreement by the parties to the contrary, there is to be implied into the agreement for adjudication the power of the adjudicator to correct an error arising from **an accidental error or omission or to clarify or remove any ambiguity in the decision** which he has reached, provided this is done within a reasonable time and without prejudicing the other party."[15]

- The power is contractual, so the parties are at liberty either to exclude the power or to limit it as they see fit (for example, in a contractual adjudication procedure). Alternatively, the adjudicator or Adjudicator Nominating Body may set out the terms of the power in the appointment agreement.
- It is for the adjudicator to decide whether there is an error or not.
- The following types of error are covered, according to *Bloor*:
  - accidental error;
  - omission;
  - clarification;
  - removal of ambiguity.

  This generally reflects section 57 of the Arbitration Act and is wide ranging. It is clear, by analogy with arbitration, that an adjudicator may correct his award to give true effect to his first thoughts and intentions, but may not change the substantive decision because he has second thoughts or intentions.
- Correction of the error must take place within a reasonable time bearing in mind the speed of the adjudication procedure. In the case of *Bloor*, this was within two and a half hours of publication of the decision; however in another case, Mr Justice Dyson, as he then was, found it at least arguable that the adjudicator had the right to correct a mistake after more than a week.[16]

---

15 *Bloor Construction (UK) Ltd* v. *Bowmer & Kirkland (London) Ltd* [2000] BLR 314, TCC at p320, His Honour Judge Toulmin.

16 *Edmund Nuttall Ltd* v. *Sevenoaks District Council* unreported 14th April 2000, TCC.

Guidance to adjudicators

1. Always check whether the parties have made an agreement excluding or limiting your right to make corrections. If they have, comply strictly with such agreement.
2. You may become aware that you have made an error, or one of the parties may raise the matter with you. In either case, consider inviting the parties (or other party) to make submissions to you.
3. When you are satisfied that you have made an error, correct it as soon as possible and notify the parties.
4. Bear in mind how much time has elapsed since you delivered the decision, and any action that the parties may have taken.
5. Remember that you have the right to decide that you have not made an error.
6. It may be appropriate (for example, in the case of decisions involving complex arithmetical calculations) to consider issuing a draft decision (in whole or part) a short time before delivering your final decision, inviting the parties to identify accidental errors.
7. Never forget your primary duty to act fairly as between the parties.

7. The Parties' Costs

The issue

There has been some uncertainty as to whether or not the adjudicator has power to decide that one party pay another party's costs.

The law

The Scheme is silent on the parties' (as opposed to the adjudicator's) costs. It was generally understood by the industry that, in the absence of any express provision in the Act or the Scheme, the adjudicator did not have the power to order one party to pay another's costs.

Some doubt was thrown on this when in one case it was decided that the Scheme gave the adjudicator the power to award costs as part of his case management powers.[17] In a subsequent case, the judge expressly disagreed with this view and this has been reinforced by a further case. In both instances, the judges clearly say that neither the Act nor the Scheme gives the adjudicator power to award costs.[18] Indeed, the second case goes further and says that a party's costs of an adjudication cannot be recovered as damages in a subsequent court action. In all the cases the judges acknowledge that the parties are free to give the power to an adjudicator by contract.

Guidance to adjudicators

It now appears clear that, in adjudication under the Scheme, each party is responsible for its own costs and you do not have the power to order one party to pay another's costs. It may be possible for you to be given that power by the parties, either by terms of the adjudication procedure, or by the parties so agreeing after the dispute has arisen.

Notes

The masculine includes the feminine.

Published by the Construction Umbrella Bodies Adjudication Task Group.

This guidance has been prepared by the British Property Federation, the Construction Confederation, the Constructors Liaison Group and the Construction Industry Council through the Construction Umbrella Bodies Adjudication Task Group in conjunction with the Construction Industry Directorate of the Department of Trade & Industry.

17 *John Cothliff Ltd* v. *Allen Build (North West) Ltd* unreported 29 July 1999, Liverpool CC, Judge Marshall Evens, (1999) CILL 1530; the Scheme, paragraph 13(h).

18 Northern Developments [Cumbria Ltd & J Nichol (2000) EWHC] 248. TCC, Judge Bowsher and *Total M and E Services Ltd* v. *ABB Building Technologies Ltd* (2000) EWHC 348, TCC, Judge Wilcox (available on www.bailii.org/ew/cases).

Appendix F

The ICE Adjudication Procedure (1997)

Contents

Acknowledgements

The Adjudication Procedure (1997) has been produced by The Institution of Civil Engineers through its Conciliation and Adjudication Advisory Panel. It was drafted by Mr Brian Totterdill with the assistance of Mr Guy Cottam.

Members of the Conciliation and Adjudication Advisory Panel are

Mr B W Totterdill BSc(Hons)(Eng) CEng FICE FIStructE FIPENZ FCIArb MAE FFB (Chairman)
Mr D Carrick FInstCES FCIArb MAE MCIPS
Mr G D G Cottam BSc(Eng) CEng FICE FIEI FCIArb MAE
Mr I G Forrest EurIng CEng MICE Dip Arb FCIArb
Mr G F Hawker TD BSc(Eng) FEng FICE CEng FIEI FIStructE MSocIS(France) MConsE FCIArb EurIng Barrister
Mr D G Loosemore CEng MICE ACIArb FInstCES
Mr T W Weddell BSc CEng DIC FICE FIStructE ACIArb

ICE Secretariat

Mrs E A Stanton (Secretary)
Mr F N Vernon BSc(Eng) CEng MICE (Technical Advisor)

Although this Procedure (approved November 1997) has been prepared by The Institution of Civil Engineers principally for use with the ICE family of Conditions of Contract it may be suitable for use with other contracts.

The Institution of Civil Engineers' ADJUDICATION PROCEDURE (1997)

1. General principles

1.1 The adjudication shall be conducted in accordance with the edition of the ICE Adjudication Procedure which is current at the date of issue of a notice in writing of intention to refer a dispute to adjudication (hereinafter called the Notice of Adjudication) and the Adjudicator shall be appointed under the Adjudicator's Agreement which forms a part of this Procedure. If a conflict arises between this Procedure and the Contract then this Procedure shall prevail.

1.2 The object of adjudication is to reach a fair, rapid and inexpensive determination of a dispute arising under the Contract and this Procedure shall be interpreted accordingly.

1.3 The Adjudicator shall be a named individual and shall act impartially.

1.4 In making a decision, the Adjudicator may take the initiative in ascertaining the facts and the law. The adjudication shall be neither an expert determination nor an arbitration but the Adjudicator may rely on his own expert knowledge and experience.

1.5 The Adjudicator's decision shall be binding until the dispute is finally determined by legal proceedings, by arbitration (if the Contract provides for arbitration or the Parties otherwise agree to arbitration) or by agreement.

1.6 The Parties shall implement the Adjudicator's decision without delay whether or not the dispute is to be referred to legal proceedings or arbitration. Payment shall be made in accordance with the payment provisions in the Contract, in the next stage payment which becomes due after the date of issue of the decision, unless otherwise directed by the Adjudicator or unless the decision is in relation to an effective notice under Section 111(4) of the Act.

2. The Notice of Adjudication

2.1 Any Party may give notice at any time of its intention to refer a dispute arising under the Contract to adjudication by giving a written Notice of Adjudication to the other Party. The Notice of Adjudication shall include:

(*a*) the details and date of the Contract between the Parties;

(*b*) the issues which the Adjudicator is being asked to decide;

(*c*) details of the nature and extent of the redress sought.

3. The appointment of the Adjudicator

3.1 Where an Adjudicator has either been named in the Contract or agreed by the Parties prior to the issue of the Notice of Adjudication the Party issuing the Notice of Adjudication shall at the same time send to the Adjudicator a copy of the Notice of Adjudication and a request for confirmation, within four days of the date of issue of the Notice of Adjudication, that the Adjudicator is able and willing to act.

3.2 Where an Adjudicator has not been so named or agreed the Party issuing the Notice of Adjudication may include with the Notice the names of one or more persons with their addresses who have agreed to act, any one of whom would be acceptable to the referring Party, for selection by the other Party. The other Party shall select and notify the referring Party and the selected Adjudicator within four days of the date of issue of the Notice of Adjudication.

3.3 If confirmation is not received under paragraph 3.1 or a selection is not made under paragraph 3.2 or the Adjudicator does not accept or is unable to act then either Party may within a further three days request the person or body named in the Contract or if none is so named The Institution of Civil Engineers to appoint the Adjudicator. Such request shall be in writing on the appropriate form of application for the appointment of an adjudicator and accompanied by a copy of the Notice of Adjudication and the appropriate fee.

3.4 The Adjudicator shall be appointed on the terms and conditions set out in the attached Adjudicator's Agreement and Schedule and shall be entitled to be paid a reasonable fee together with his expenses. The Parties shall sign the agreement within seven days of being requested to do so.

3.5 If for any reason whatsoever the Adjudicator is unable to act, either Party may require the appointment of a replacement adjudicator in accordance with the procedure in paragraph 3.3.

4. Referral

4.1 The referring Party shall within two days of receipt of confirmation under 3.1, or notification of selection under 3.2, or appointment under 3.3 send to the Adjudicator, with a copy to the other Party, a full statement of his case which should include:

(*a*) a copy of the Notice of Adjudication;

(*b*) a copy of any adjudication provision in the Contract, and

(*c*) the information upon which he relies, including supporting documents.

4.2 The date of referral of the dispute to adjudication shall be the date upon which the Adjudicator receives the documents referred to in paragraph 4.1. The Adjudicator shall notify the Parties forthwith of that date.

5. Conduct of the adjudication

5.1 The Adjudicator shall reach his decision within 28 days of referral, or such longer period as is agreed by the Parties after the dispute has been referred. The period of 28 days may be extended by up to 14 days with the consent of the referring Party.

5.2 The Adjudicator shall determine the matters set out in the Notice of Adjudication, together with any other matters which the Parties and the Adjudicator agree should be within the scope of the adjudication.

5.3 The Adjudicator may open up, review and revise any decision (other than that of an adjudicator unless agreed by the Parties), opinion, instruction, direction, certificate or valuation made under or in connection with the Contract and which is relevant to the dispute. He may order the payment of a sum of money, or other redress but no decision of the Adjudicator shall affect the freedom of the Parties to vary the terms of the Contract or the Engineer or other authorised person to vary the Works in accordance with the Contract.

5.4 The other Party may submit his response to the statement under paragraph 4.1 within 14 days of referral. The period of response may be extended by agreement between the Parties and the Adjudicator.

5.5 The Adjudicator shall have complete discretion as to how to conduct the adjudication, and shall establish the procedure and timetable, subject to any limitation that there may be in the Contract or the Act. He shall not be required to observe any rule of evidence, procedure or otherwise, of any court. Without prejudice to the generality of these powers, he may:

(*a*) ask for further written information;

(*b*) meet and question the Parties and their representatives;

(*c*) visit the site;

(*d*) request the production of documents or the attendance of people whom he considers could assist;

(*e*) set times for (a)–(d) and similar activities;

(*f*) proceed with the adjudication and reach a decision even if a Party fails:

  (*i*) to provide information;

  (*ii*) to attend a meeting;

  (*iii*) to take any other action requested by the Adjudicator;

(*g*) issue such further directions as he considers to be appropriate.

5.6 The Adjudicator may obtain legal or technical advice having first notified the Parties of his intention.

5.7 Any Party may at any time ask that additional Parties shall be joined in the Adjudication. Joinder of additional Parties shall be subject to the agreement of the Adjudicator and the existing and additional Parties. An additional Party shall have the same rights and obligations as the other Parties, unless otherwise agreed by the Adjudicator and the Parties.

6. The Decision

6.1 The Adjudicator shall reach his decision and so notify the Parties within the time limits in paragraph 5.1 and may reach a decision on different aspects of the dispute at different times. He shall not be required to give reasons.

6.2 The Adjudicator may in any decision direct the payment of such simple or compound interest at such rate and between such dates or events as he considers appropriate.

6.3 Should the Adjudicator fail to reach his decision and notify the Parties in the due time either Party may give seven days' notice of its intention to refer the dispute to a replacement adjudicator appointed in accordance with the procedures in paragraph 3.3.

6.4 If the Adjudicator fails to reach and notify his decision in due time but does so before the dispute has been referred to a replacement adjudicator under paragraph 6.3 his decision shall still be effective.

If the Parties are not so notified then the decision shall be of no effect and the Adjudicator shall not be entitled to any fees or expenses but the Parties shall be responsible for the fees and expenses of any legal or technical adviser appointed under paragraph 5.6 subject to the Parties having received such advice.

6.5 The Parties shall bear their own costs and expenses incurred in the adjudication. The Parties shall be jointly and severally responsible for the Adjudicator's fees and expenses, including those of any legal or technical adviser appointed under paragraph 5.6, but in his decision the Adjudicator may direct a Party to pay all or part of his fees and expenses. If he makes no such direction the Parties shall pay them in equal shares.

6.6 At any time until seven days before the Adjudicator is due to reach his decision, he may give notice to the Parties that he will deliver it only on full payment of his fees and expenses. Any Party may then pay these costs in order to obtain the decision and recover the other Party's share of the costs in accordance with paragraph 6.5 as a debt due.

6.7 The Parties shall be entitled to the relief and remedies set out in the decision and to seek summary enforcement thereof, regardless of whether the dispute is to be referred to legal proceedings or arbitration. No issue decided by an adjudicator may subsequently be laid before another adjudicator unless so agreed by the Parties.

6.8 In the event that the dispute is referred to legal proceedings or arbitration, the Adjudicator's decision shall not inhibit the court or arbitrator from determining the Parties' rights or obligations anew.

6.9 The Adjudicator may on his own initiative, or at the request of either Party, correct a decision so as to remove any clerical mistake, error or ambiguity provided that the initiative is taken, or the request is made within 14 days of the notification of the decision to the Parties. The Adjudicator shall make his corrections within seven days of any request by a Party.

7. Miscellaneous provisions

7.1 Unless the Parties agree, the Adjudicator shall not be appointed arbitrator in any subsequent arbitration between the Parties under the Contract. No Party may call the Adjudicator as a witness in any legal proceedings or arbitration concerning the subject matter of the adjudication.

7.2 The Adjudicator shall not be liable for anything done or omitted in the discharge or purported discharge of his functions as Adjudicator unless the act or omission is in bad faith, and any employee or agent of the Adjudicator shall be similarly protected from liability. The Parties shall save harmless and indemnify the Adjudicator and any employee or agent of the Adjudicator against all claims by third parties and in respect of this shall be jointly and severally liable.

7.3 Neither The Institution of Civil Engineers nor its servants or agents shall be liable to any Party for any act, omission or misconduct in connection with any appointment made or any adjudication conducted under this Procedure.

7.4 All notices shall be sent by recorded delivery to the address stated in the Contract for service of notices, or if none, the principal place of business or registered office (in the case of a company). Any agreement required by this Procedure shall be evidenced in writing.

7.5 This Procedure shall be interpreted in accordance with the law of the Contract.

8. Definitions

8.1 (a) The "Act" means the Housing Grants, Construction and Regeneration Act 1996.

(*b*) The "Adjudicator" means the person named as such in the Contract or appointed in accordance with this Procedure.

(*c*) "Contract" means the contract or the agreement between the Parties which contains the provision for adjudication.

(*d*) "Party" means a Party to the Contract and references to either Party or the other Party or Parties shall include any additional Party or Parties joined in accordance with this Procedure.

9. Application to particular contracts

9.1 When this Procedure is used with The Institution of Civil Engineers' Agreement for Consultancy Work in Respect of Domestic or Small Works the Adjudicator may determine any dispute in connection with or arising out of the Contract.

A copy of the appropriate form for applying for the selection/appointment of an Adjudicator by The Institution of Civil Engineers may be obtained from:

The Dispute Administration Service, The Institution of Civil Engineers, One Great George Street, Westminster, LONDON SW1P 3AA

Telephone: +44 (0)20 7222 7722
Facsimile: +44 (0)20 7222 1403.

Adjudicator's agreement

THIS AGREEMENT is made on the .... day of 20
between (the first Party):
of:

and (the second Party):
of:

and (where there is a third Party):
of:

(hereinafter called "the Parties") of the one part and:
of:

(hereinafter called "the Adjudicator") of the other part.

Disputes or differences may arise/have arisen* between the Parties under a Contract dated .... and
known as:

and these disputes or differences shall be/have been* referred to adjudication in accordance with The Institution of Civil Engineers' Adjudication Procedure (1997) (hereinafter called "the Procedure") and the Adjudicator has been requested to act.
* Delete as necessary

IT IS NOW AGREED as follows:

1. The rights and obligations of the Adjudicator and the Parties shall be as set out in the Procedure.
2. The Adjudicator hereby accepts the appointment and agrees to conduct the adjudication in accordance with the Procedure.
3. The Parties bind themselves jointly and severally to pay the Adjudicator's fees and expenses in accordance with the Procedure as set out in the attached Schedule.
4. The Parties and the Adjudicator shall at all times maintain the confidentiality of the adjudication and shall endeavour to ensure that anyone acting on their behalf or through them will do likewise, save with the consent of the other Parties which consent shall not be unreasonably refused.
5. The Adjudicator shall inform the Parties if he intends to destroy the documents which have been sent to him in relation to the adjudication and he shall retain documents for a further period at the request of either Party.

Signed on behalf of:

First Party:

Name:

Signature:

Date:

Second Party:

Name:

Signature:

Date:

Third Party (where there is a third Party)**:**

Name:

Signature:

Date:

Adjudicator:

Name:

Signature:

Date:

SCHEDULE to the ADJUDICATOR'S AGREEMENT

1. The Adjudicator shall be paid at the hourly rate of £ in respect of all time spent upon, or in connection with, the adjudication including time spent travelling.
2. The Adjudicator shall be reimbursed in respect of all disbursements properly made including, but not restricted to:
   (*a*) Printing, reproduction and purchase of documents, drawings, maps, records and photographs.
   (*b*) Telegrams, telex, faxes, and telephone calls.
   (*c*) Postage and similar delivery charges.
   (*d*) Travelling, hotel expenses and other similar disbursements.
   (*e*) Room charges.
   (*f*) Charges for legal or technical advice obtained in accordance with the Procedure.
3. The Adjudicator shall be paid an appointment fee of £ . This fee shall become payable in equal amounts by each Party within 14 days of the appointment of the Adjudicator. This fee will be deducted from the final statement of any sums which shall become payable under item 1 and/or item 2 of this Schedule. If the final statement is less than the appointment fee the balance shall be refunded to the Parties.
4. The Adjudicator is/is not* currently registered for VAT.
5. Where the Adjudicator is registered for VAT it shall be charged additionally in accordance with the rates current at the date of invoice.
6. All payments, other than the appointment fee (item 3) shall become due seven days after receipt of invoice, thereafter interest shall be payable at 5% per annum above the Bank of England base rate for every day the amount remains outstanding.

* Delete as necessary

Notice of intention to refer a dispute to adjudication

To: *(Name of Responding Party)*
(Address of Responding Party)

Date:

Dear Sir

NOTICE OF ADJUDICATION
(Reference)

We consider that a dispute or difference has arisen between us under:

(a) *(Contract Name:)*
(Contract Date:)
(Contract Location:)

(b) Included as Appendix A are the issues which the Adjudicator is being asked to decide

(c) Included as Appendix B are the details of the nature and extent of the redress sought

We now give notice that we require this dispute or difference to be referred to adjudication.

To act as Adjudicator the following person(s)

has been named in the contract and has been sent a copy of this Notice [1]
has/have agreed to act and is/are proposed for your consideration [2]

Name(s) of Adjudicator(s)
Address

Within 4 days of the date of this Notice

the named Adjudicator has been requested to confirm he is able and willing to act[1]
you are required to agree in writing to the appointment[2]

failing which within a further 3 days The Institution of Civil Engineers will be requested to select/appoint the Adjudicator.

Yours faithfully

For and on behalf of
(Referring Party)

1 and 2 Select either option 1 or option 2

ICE
Dispute
Administration
Service

Application for the selection/appointment of an Adjudicator

Dispute Administration Service
The Institution of Civil Engineers
One Great George Street
Westminster
London SW1P 3AA

Date ....

The dispute and/or difference described in the attached Notice of Adjudication Reference _______ has arisen and since an Adjudicator has not been agreed between the Parties I/we hereby apply to you to select/appoint an Adjudicator who has experience in:

Civil Engineering   Building   Process Engineering *

I/we agree to meet all the reasonable costs incurred by the person selected/appointed by you if I/we am/are not entitled to make this application in accordance with the agreement between the parties.

I/we enclose a cheque for £_______ + VAT in respect of the charge made by the Institution towards the administrative cost in connection with this application.

Yours faithfully

For and behalf of: ....

Name: ....

Address: ....

....

*Please delete whichever is not appropriate

Notes

1. *The Institution will make a selection/appointment upon the application of any person using this form. The validitity of the application will not be investigated.*
2. *In making this application the Applicant undertakes to meet the reasonable charges of the person selected/appointed by the Institution pursuant to this application should the adjudication not proceed.*
3. *If the validity of the application is challenged then the person appointed and the parties involved must resolve the challenge.*
6. *Neither the Institution nor its servants or agents nor its members shall be liable to any party for any act omission or misconduct in connection with any selection or appointment made or any adjudication conducted in consequence of this application.*

Notice of adjudication and fee must accompany this form

ICE
Dispute
Administration
Service

Selection/appointment of an Adjudicator by the Institution of Civil Engineers

To (Name and Address of Applicant): ....

I hereby select/appoint: ....

of ....

Adjudicator for the dispute or difference described in the Notice of Adjudication attached to this application.

Dated: ....

On behalf of the Institution of Civil Engineers

Copies for information:

The Adjudicator

The Parties (other than the Applicant)

Appendix G

Published Adjudication Procedures and Adjudicator Nominating Bodies

Publishers of Adjudication Procedures

The following Adjudication Procedures have been published and are said to comply with the Housing Grants, Construction and Regeneration Act 1996. Users should check for themselves that they do comply.

1. **The Institution of Civil Engineers' Adjudication Procedure (1997)**
   The Institution of Civil Engineers
   1 Great George Street
   Westminster
   London SW1P 3AA
   Tel: 020 7665 2214
   Fax: 020 7222 1403
   Website: www.ice.org.uk

2. **TeCSA Adjudication Rules**
   The Technology and Construction Solicitors Association
   Peter Rees, Chairman
   C/o Norton Rose
   Kempson House
   Camomile Street
   London EC3A 7AN
   Tel: 020 7283 6000
   Fax: 020 7283 6500
   Website: www.tecsa.org.uk

3. **CEDR Model Rules for Adjudication**
   Centre for Dispute Resolution
   Princes House
   95 Gresham Street
   London EC2V 7NA

4. **CIC Model Adjudication Procedures: First Edition**
   The Construction Industry Council (CIC)
   26 Store Street
   London WC1E 7BT
   Tel: 020 7637 8692
   Fax: 020 7580 6140
   Website: www.cic.org.uk

5. **IChemE – The Institution of Chemical Engineers**
   Davis Building,
   165–189 Railway Terrace
   Rugby
   Warwickshire CV21 3HQ
   Tel: 01788 578214
   Fax: 01788 560833
   Website: www.icheme.org/learning

Adjudicator Nominating Bodies

The following bodies have stated that they are prepared to select Adjudicators to act in connection with the Scheme.

1. **ICE – The Institution of Civil Engineers**
   1 Great George Street
   Westminster
   London SW1P 3AA
   Tel: 020 7665 2214
   Fax: 020 7222 1403
   Website: www.ice.org.uk
   E-mail: drick.vernon@ice.org.uk; peter.madigan@ice.org.uk

2. **TeCSA – The Technology and Construction Solicitors Association**
   Peter Rees, Chairman
   C/o Norton Rose
   Kempson House
   Camomile Street
   London EC3A 7AN
   Tel: 020 7283 6000
   Fax: 020 7283 6500
   Website: www.tecsa.org.uk

3. **CEDR – Centre for Dispute Resolution**
   Princes House
   95 Gresham Street
   London EC2V 7NA
   Tel: 0870 333 1600
   Fax: 020 7600 0501
   E-mail: mediate@cedr.co.uk

4. **RICS – The Royal Institution of Chartered Surveyors**
   RICS Contact Centre
   Surveyor Court
   Westwood Way
   Coventry CV4 8JE
   Tel: 0870 333 1600
   Website: www.rics.org.uk

5. **CIOB – The Chartered Institute of Building**
   Englemere
   Kings Ride
   Ascot
   Berkshire SL5 7TB
   Tel: 01344 630745
   Fax: 01344 630713
   Website: www.ciob.org.uk
   Email: cking@ciob.org.uk

6. **CIArb – The Chartered Institute of Arbitrators**
   Dispute Resolution Services
   International Arbtiration Centre
   12 Bloomsbury Square
   London WC1A 2LP
   Tel: 020 7421 7444
   Fax: 020 7404 4023
   Website: www.arbitrators.org

7. **RIBA – The Royal Institute of British Architects**
   66 Portland Place
   London W1N 4AD
   Tel: 020 7580 5533
   Fax: 020 7255 1541
   Email: admin@inst.riba.org

Appendix H

Index to Paragraphs in the Scheme

Part I. Adjudication

Part II. Payment

Paragraph	Reference	Page
3		125
4		126
5		128
6		129
7		126
8	(1)	126
	(2)	126
9		130
10		131
11		135
12		122, 123, 124, 125, 126
20		120

Index